We Are The Nibiruans

Return of the 12th Planet

We Are The Nibiruans

Return of the 12th Planet

BOOK ONE

By Jelaila Starr

The Nibiruan Council
www.nibiruancouncil.com

Library of Congress Cataloging-in-Publication Data
Starr, Jelaila, 1956
We Are The Nibiruans, Return of the 12th Planet / by Jelaila Starr.
p. cm
"Book one."
ISBN 0-9656657-0-4

1. Spirit writings.
2. Life on other planets--Miscellanea.
1. Title.

BF1311.L43573 1999   99-32650

133.9'3--dc21   CIP

Cover Artwork: Phyllis Furphy

Phyllis is a uniquely talented artist and starseed. She draws the most
exquisite portraits of Guides, Angels, Starseed Fathers and people on
other dimensions as well as pictures of other realms. You may reach her
at 916.441.4412.

4th Edition

Printed in the United States of America.

Address all inquiries to:

The Nibiruan Council
Email: info@nibiruancouncil.com
Website: www.NibiruanCouncil.com

In Memory of

Joscelyn

for preparing the way for me.

Dedication

To Devin, my 9D brother and Guide

Thank you, my love, for never giving up or losing faith in me. I know I was a real challenge at times. Your wisdom and straight-forward manner has never lost my respect.

You taught me how to stop looking up to you and begin walking beside you. My feelings, thoughts, and memories of you and me would be a book in itself, wouldn't it, love?

I eagerly await the day when we will again be able to walk, arm-in-arm, down to our favorite spot by the lake where we live. Let's re-create our paradise of Avyon on Earth for all mankind to enjoy.

Promise me Devin, that when this mission is over, we will take that walk.

Jelaila

Table of Contents

Part Three — DNA Recoding, Reconnection, & Activation

Acknowledgments

My Parents, Ed and Modean Kelley — Thank you for the lessons of my childhood. Without you and them, I couldn't be who I am now.

My former husband Rick – For you I have a special love. You taught me the lesson of tenacity and you stood beside me when the going got tough and for those precious gifts, I thank you.

Bob and Tomi Wellmaker – The two of you were like my second parents. I want to thank you, Tomi, for the temperament training. You are such a gracious lady. Bob, I feel honored to have been taught by you. You were one of my greatest teachers in this dimension. You taught me to believe in myself and inspired me to stretch from one comfort zone to another. I thank you and will always be grateful.

My friend and business partner, Marla Westrup – We have covered many miles together and shed many tears. For me, it has been all worth it. You would still be the one I would go through Devin's training with, if I had to do it all over again. Thank you for your laughter during the hard times and your friendship. You have been an inspiration.

My friends Bobbie Martin, Pat Marshall, and Ann Brewer – Thank you for your love, friendship, and emotional support during and after I walked in. I love you all.

My friends, Shasarai, Versarai, and Malarai – Thank you for your support and belief in all we are doing. Your love and friendship mean a lot to me.

Barry Becker and James – See, I kept my promise. I thank you both, from the bottom of my heart, for teaching me how to have fun and be a kid again. I am eternally grateful.

To my most challenging and loving lesson mate, Chris Becker –
Thank you, Chris, for your love and faith in me. You taught me to
love myself by teaching me boundaries. You taught me to respect
myself by taking back my power. You taught me the meaning of true
intimate love and for that I will always love you. This was not to be
our happily-ever-after lifetime. Maybe the next. I wish for you all
the happiness this life has to offer. I wish you love.

To my baby girl, Danielle –Loving you has been my greatest joy. I
walked in when you were four, and instead of pushing me away, you
completely opened your heart to let me in. Thank you for your
acceptance, for allowing me to experience with you that special kind
of love only known to mothers and their daughters.

To my Life Mate, wherever you may be – My faith in your reality
keeps me moving on.

Preface

by Alystar

Keeper of the Records for the Galactic Federation

Greetings of Love to you, mankind of Earth. I am Alystar. I have been asked by Devin, of the Galactic Federation's 9D Nibiruan Council, to write the preface for this most inspiring and Earth-shaking manuscript. As one who maintains the records for the Federation, I am intimately aware of the history of all souls in this galaxy. I would like to introduce one of those souls to you now – Jelaila.

Jelaila has been chosen to head the work on Earth for the Nibiruan Council. She has a long history of performing this service in other galaxies, universes, and dimensions. Therefore, we have chosen her based on her record.

I would also like to mention that we have prepared her for every eventuality and you can depend on her to perform her duties of service to you. After four years of Earth-time training, we are ready to place Jelaila into service and make her publicly known. The Nibiruan Council has great faith in her and, therefore, so can you.

With love and joy we present to you our messenger. May she honor us and Earth with her service.

Blessings and Love,

Alystar
The Galactic Federation's Keeper of the Records

Introduction
by Versarai

This book is a handbook for ascension of the mind and, eventually, the physical body; a process we can choose to go through purposefully and consciously. It's not happenstance or serendipity. People don't default through it. It's that proverbial eye of the needle, and this book has guided many through the eye.

I have been working personally with Jelaila since 1996, training and helping others as a coach of what we have come to call the RRA Process, short for Recoding, Reconnection, and Activation as presented in this volume.

I, Versarai, was one of the early ones to apply the Formula of Compassion (see Part 3) over and over, to peel layer after layer off of my emotional blocks. Jelaila worked with me in numerous sessions as I hit my emotional walls, getting deep into the hidden and denied parts of myself. Her patient and persistent guidance with the steps in this volume helped me finally re-emerge into my own multi-dimensionality, complete with functioning psychic glands and compassion for myself and others.

In this book are the basics for completing your chosen life lessons. By using these basics, applying them everyday, making them a part of their consciousness, I have seen many people go into their greatest fears; deeply feeling and flagrantly expressing them, to gain the higher perspective. Their reward? Release from the bondage of their fears that has ruled their lives for many eons.

This process works and everyone can do this. It is simple, yet often not easy, for you must go into your emotional stuff. Many people became multi-dimensional over a short period of time doing just this process on a daily basis. They realized their higher potential as their psychic glands began to work again. This is the higher dimensional book that will help you through your process if you choose.

The three parts of this book are set up to:

1. Satisfy the reader's need to know who is sharing this information and how she came to have this information, given in Part 1;

2. Satisfy the reader's curiosity and perspective, in Part 2, with an updated universal history to 2000 B.C. as presented from Anu through Jelaila. This process involved Jelaila thoroughly researching a long list of reference materials. Then Anu used Jelaila's memory banks to help the essential pieces flow into a historical order. The universal history helps you, the reader, open to the reasoning behind the progressive development and evolution of a planetary system, and the use of a soul evolution game—The Polarity integration Game.

3. Share the tools for emerging from this soul evolution game. These are the final pieces to finally graduate, channeled from Joysia and given in Part III. These include the road map of the nine levels of changes all of humanity will eventually grow through, and you can choose to do proactively. These changes result in Accelerated DNA Recoding, Reconnection, and Activation. The vehicle to navigate these changes is the Formula of Compassion.

Six supplemental tools have emerged as we have assisted many with their recoding process: "The Compassion Key," "The Soul/Ego Partnership," "The Open Door Key," and "The Hold onto Nothing Key." These are available in audiotape and booklet form.

A network of coaches has been trained to assist those who desire more help with using these Ascension tools. They are ready to assist 'recoders' individually and in workshop settings. See the Nibiruan Council website for more details: www.NibiruanCouncil.com.

I leave you, the reader, with the option to read further. See for yourself how and why this works, and why it is the key to ascension in this time of accelerated change.

My hope is, having applied this knowledge and gained the benefits of living life compassionately and multi-dimensionally, that the template created here will open wide the roadway for you and many others to quickly and easily do this for yourselves and all of humankind.

Congratulations on getting to this point. As this book is in your hands, you are ready to begin your own emergence from the fear and pain of third dimensional Earth into your multi-dimensional life. This is your handbook to know the way and successfully navigate the path. May you enjoy your journey.

Part One

JOSCELYN

By Jelaila

1

Joscelyn

The Early Years

Unless you are a walk-in, you can only imagine how it feels to be writing about a person you thought for so long was you, only to find out that you are someone different—one of the oversouls.

Joscelyn was born during a blizzard, in Toledo, Ohio, on December 27, 1956, at 7:10 p.m. Her three older brothers had the chicken pox, so her mother had to drive herself to the hospital. Joscelyn always found it strange that her father could not be there for the delivery. She felt that there was something very different about the circumstances surrounding her birth. It seemed as though the universe had arranged things so her father could not be there. Years later, after I walked in, I would begin to understand why.

Joscelyn's father was an electrical engineer, and her mother a homemaker. Theirs was an average middle-class family. She never wanted for anything important.

Joscelyn grew up as the only girl in a family of six boisterous kids and never felt like she was really a part of the family. She spent most of her time in the woods. She had a special fondness for tree climbing and bird watching. Many were the times she wished she could fly like a bird and talk their language. A couple of times she tried to fly and barely escaped without a broken bone. Many a time she would sit in her favorite tree and wonder why she couldn't lift off and fly with those she considered her family. At night she would look at the stars and, once again, feel the great surge of longing for a home she knew existed, but could not find. The feelings of abandonment would well up inside her and bring on her short periods of depression.

Joscelyn was a bright, inquisitive, strong-willed child. She had a silly sense of humor and would entertain her father with theatrical skits she and her brother, Keith, would create. She was the clown of the family and yet, the loner. What little sense of belonging she developed came through her role as the family jester. She was slender, with blonde hair that would later turn reddish brown. The only thing different in her coloring was her eyes. No one in the family had those eyes. They were shaped like cat's eyes. The irises were golden brown around the center with gold flecks in a green background. Many were the times her mother would comment on her eyes, but the comments always had an edge to them.

Her family moved about every three years, due to the nature of her father's work. In 1973 they settled in Birmingham, Alabama, after moving from Nashville, Tennessee, with a few shorter moves in between. The periodic moves provided an excellent opportunity for her to learn flexibility and to practice her social skills. She made friends easily, but never let them get too close. In school she tended to float amongst the different groups, never really becoming an integral part of any one, yet friendly with all. Joscelyn struggled academically. She made good grades, but never felt inspired to do her best unless it involved singing, speaking, or other rightbrained activities.

After she became too old for tree climbing, she spent her time daydreaming and reading. You could always find her with her nose in a book about cats, horses, birds, and fairytales. She would go to the library and check out a stack of five or six books, which she would read within a week and return. In a few years, her interests would turn to European history, with an emphasis on the monarchies of Europe. This trend would be followed by a passionate interest in religious history with an emphasis on the twelve tribes of Israel and Egyptian royalty.

As a child Joscelyn didn't have a thought about what she would be when she grew up. In fact, I don't find any references in her memory tapes concerning a desire to be anything. It was almost as though she knew she wouldn't be around for the duration, so she lived her young life from day to day, reading and playing. It wasn't until she entered high school that she chose a career—singing. She wanted to be a singer and play the piano, so she daydreamed the hours away thinking about it. It was one of the few dreams that Joscelyn would never realize. I can understand her burning desire to be a singer, now that I know of her Victorian lifetime as a well-known singer.

Joscelyn grew up in a family whose life revolved around a religion which I call a cult, for lack of a better term. They belonged to the Worldwide Church of God and were taught strict obedience to the Church and to God.

The Church believed in a form of isolationism. You could socialize outside of the Church, but dating and marrying outside the Church were heavily frowned upon. The Church was a blessing in disguise for Joscelyn. Because of its teachings she would learn a lot about Biblical history, which she would need as a foundation for the research I, Jelaila, would begin, after I walked in, in 1992.

It also provided her with a strong belief system based on fear, which I, Jelaila, would later release. I would understand fear and move beyond it by working through and integrating her fears. To accomplish this, I did many years of research on the religious systems on Earth. I discovered how they were set up and managed, and that they were based on fear. Once I realized this I began my steady ascent out of that system. My ascent out of the system mirrors the ascent of Humankind out of the Piscean Age of the parent/child religious system where the "Church" is the parent and the "followers" are the children. We are now entering the Aquarian age where each individual realizes as I did that they do not need a "church" or parent to find God.

By working through her life contract and completing it, I, Jelaila, gained sufficient experience and compassion for mankind on Earth.

First Psychic Experience

When Joscelyn was twenty-one, she went to her first psychic, out of curiosity. Her name was Mrs. Smith. Joscelyn's friend, Luanne, had told her about Mrs. Smith, a long-time friend of Luanne's family. Luanne's family had long been gifted with psychic abilities. Her brother, John, saw dead family members, and her younger twin brother and sister were telepathic. The twins could sit on opposite ends of the room and read what each had written on a piece of paper. Joscelyn had witnessed this many times and found it fascinating.

Going to Mrs. Smith was a big step for Joscelyn. The Church banned members from seeking out these people, saying they were demon possessed. Joscelyn, being the renegade she was, felt this made it even more intriguing and went anyway. This reading was the first in a chain of events that would leave clues as to Joscelyn's real identity and parentage.

Mrs. Smith told Joscelyn, among other things, that her father was not her real father. This was a source of great distress for her. That would mean that her mother, at some time during her marriage, had had an affair, yet Joscelyn looked more like her father's family than her mother's family. This left her completely bewildered. It never occurred to her to go back to Mrs. Smith and ask for more information on her parentage. Instead she chose to keep it from her mother and held it inside for many years. Somehow, deep inside her, she knew Mrs. Smith was right but she also knew her mother had not been unfaithful. For her, it was another piece to add to her growing number of unanswered questions about her life.

Mrs. Smith's information added to Joscelyn's feelings of alienation. She left home, Birmingham, Alabama, at the age of twenty two, and she didn't just move across town—she moved 500 miles away to Houston, Texas! Unconsciously, she knew she needed to break away from the hold her family and the Church had on her.

In Houston, she learned about life and over-worked her guardian angel. Houston was experiencing the oil boom of 1979 through 1981. It was a great time to be young, pretty, and on your own in a big, wild, city like Houston. She was too naive to know she should have been scared.

Joscelyn met Mark, an attorney, through her roommate, Joelyn. She went to work for him for a brief period of time, but wasn't much for sitting behind a desk. So she decided to go into restaurant work. She made good money working in the higher-end restaurants, usually around $100 a night in tips—good money for a girl of twenty-two, in that economy. It allowed her to indulge her love of pretty, stylish clothes and nice cars. She always had a beautifully decorated apartment. In fact, the leasing office often used her apartment to show to prospective tenants.

Joscelyn dated many men, but none for very long, preferring to keep them from getting too close. She usually chose men with brains and power over men with looks. Joscelyn had many adopted mothers and fathers—often, the parents of her boyfriends or her girlfriends. People liked to take her under their wing. Maybe they could sense her loneliness and need for love and acceptance.

At twenty-four she married for the first time. Bob, her husband, was in the Church. During her years in Houston Joscelyn longed to be accepted by her family—she had become the black sheep before the age of twelve. She felt marrying Bob would give her this acceptance because her parents wanted her baptized and married. (One of the Church requirements was that members must be baptized before marriage.)

So she took the baptismal plunge, married Bob, and moved to Longview, Texas. The date was January 10, 1981. It was a marriage based on passion and a need for acceptance. She knew she was making a mistake but, true to her nature, she told herself she would just make it work. Little did she know that she was out of her league.

Bob was an alcoholic and a wife beater. Twice, she experienced his drunken rages and was ready to leave, but the Church said she could not divorce him. The final straw came in the form of a miscarriage, brought on by the tremendous stress, among other things, of the relationship. Joscelyn decided to go against Church policy. She called the minister and told him she wasn't going to stay and proceeded to get the divorce. They had been married only a year.

After the divorce, Joscelyn went to work for the Wellmakers. They had a business just west of Big Sandy, in a little town called Hawkins, twenty minutes from Longview. The Wellmakers were friends of her family. They had known each other since she was fourteen. The parents, Bob and Tommie, had two wonderful children, Jeff and Amy, who were just a few years younger than Joscelyn.

Her family had met the Wellmakers when they moved in next door to them in Ardmore, Oklahoma, in 1970. Mr. Wellmaker was a wealthy deacon in the Church—a consummate entrepreneur with a brilliant mind and the Midas Touch. He could read, motivate, and empower people better than anyone Joscelyn had ever known.

The Wellmakers owned a marketing business at that time and had expanded it across the country. Bob asked Joscelyn to come work for him. He told her he would send her to school to learn computers. So, Joscelyn enrolled at Tyler Junior College, in Tyler, Texas. After about two weeks of three-hour-a-night classes, she began skipping out early. On any given night you could find her driving around Tyler, listening to Michael Jackson and drinking a Coke. Finally, she went to Bob and said she wasn't cut out for school or computers, so he put her in the marketing department, which suited her better. This resulted in a transfer to Kansas City, Missouri, to open a new office in April of 1983.

The Wellmakers lived the rich life of big houses, expensive cars and even private planes, and, as a result, so did Joscelyn. She lived in a company condominium and drove a company car. She was the official meeting planner for the company and the personal assistant of Mr. Wellmaker. They were very generous, to say the least. They "adopted" her as they had her brother, Tim, years before, and they gave her a great education in the world of entrepreneurism and marketing. Mr. Wellmaker, especially, took her under his wing and taught her how to motivate, inspire, and lead people. It would prove to be invaluable training for the future.

Joscelyn resigned from her position as meeting planner with the Wellmakers after she met Wes Blake, another self-made millionaire, in January of 1984. She flew in February to Phoenix, Arizona, to meet him, and they spent the next month in the warm sunshine of Arizona, where he was vacationing, indulging their love of antiquing and sight-seeing.

They lived together for nearly a year. Wes also taught her much about business. In 1985, they split up. Joscelyn moved to her own apartment and took a job as executive secretary to the president of a large trucking and warehousing firm.

The break-up of this relationship was one of the most painful experiences of her life. On the other hand, it gave her an opportunity to balance karma carried over from her lifetime in the court of Henry VIII. In that life she had jilted Wes for someone with more power and influence. This time around she had been the jilted one! The effects of balancing that karma were emotionally and physically devastating. She had never experienced such powerful emotions. It became so bad, she had to call her mother to come to Kansas City to stay with her. She lost fifteen pounds from not being able to eat, and looked like a skeleton. When her mother arrived she was shocked to see her daughter in such a state.

Joscelyn also lost her job because she couldn't go more than a hour or two without having to run to the bathroom to cry. Although her employer had been very understanding and patient, even he had his limits. To Joscelyn, the only possible solution was to move to another part of the city. This too, was a blessing in disguise. Through the intense emotions, Joscelyn learned what it was like to deeply feel. Before this, she had felt little emotion for anything.

After her break-up with Wes, in the fall of 1985, Joscelyn moved to a part of Kansas City called Midtown. She had been living in an area called "North of the River" in the northern part of Kansas City. Midtown was an eclectic part of the city and provided Joscelyn with her first experience of real life. Rich, poor, gays, musicians, dancers, college students, and drug addicts were her neighbors. What a great opportunity to learn acceptance for people from every walk of life.

In the spring of 1986 she met Rick, who was sweet, quiet, and handsome. She liked him because of his easy-going manner and total lack of pretense. What you see is what you get with Rick. With both being strong willed and stubborn, they argued quite often. Joscelyn even threw him out of her apartment three times during their courtship, but he kept coming back. Tenacity was one of Rick's strong suits.

Her landlords at the time, a gay couple she loved dearly, had a bet going that she would marry him. They used to watch over her and take care of her cats when she was gone. No one could have asked for better landlords and surrogate fathers. They would all get together on Sunday mornings and sit out on the balcony and watch the street people go by. The apartment was a fourplex, built at the turn of the century, and right off Main street, so there was always a lot going on. It was a great place to live if you liked observing people from all levels of society.

In June, Rick and Joscelyn decided to move in together. Joscelyn had started a company called Team Marketing Concepts with a partner, Dave. They were ready to begin expanding into St. Louis, Missouri, and Joscelyn knew she would need someone to take care of her three cats. Also, dating was a challenge, since Olathe was 35 minutes from Midtown. Sharing a home together seemed the most logical solution.

Rick moved out of his apartment and Joscelyn moved out of hers, or rather Rick's mother moved Joscelyn out of hers. The following weekend Joscelyn found out she was carrying their child. Actually Rick was the one who discovered it. They were lying on the sofa on Sunday afternoon and Rick had his arm around her as they watched TV. He casually said during the commercial break, "Did you know that you're pregnant?" Rick is very psychic. Both were excited and decided the best thing to do for all concerned was to tie the knot. Joscelyn and Rick married on July 2, 1987. They moved to a house in Olathe, a suburb of Kansas City, on the Kansas side of the state line.

Rick had a carpet and design business in Olathe, Kansas that was doing quite well, financially. Joscelyn could not say that she was in love with Rick but, she loved him and knew she had something to fulfill and complete with him. She loved Rick for the sweet, lovable person he was, and after her first marriage, based on passion, she felt maybe one based on friendship would work better. The fact that she became pregnant was a real source of mystery to her, since she had tried for over a year with her previous fiancee, Wes, to have a child. Wes had wanted a child very much and Joscelyn's apparent infertility was a source of conflict in their relationship.

It had been six years since her first divorce, and her miscarriage during that marriage. She had believed that she could not conceive and was overjoyed at the prospect of being a mother. I have memory tapes of her talking with her unborn child. She knew it would be a girl. Rick and his family thought she was going through some sort of prenatal insanity because Joscelyn kept insisting that she was having conversations with the baby and that the baby was a girl who wanted to be named Danielle. Months later her prediction that the baby was a girl was proven correct. For the next two years she worked on being a good wife and mother. Her spiritual studies were put on the back burner until after her divorce in 1989.

By December of 1988, when Danielle was nine months old, Joscelyn knew the marriage was not going to survive. Joscelyn had begun having nightmares about being left to die in the forthcoming "End Times" due to her unworthiness. The Church taught that a third of the world's population would die by famine, a third by pestilence, and a third by war. Only the "Elect" would survive by being taken to a place of safety. To be among the "Elect," one would have to be a baptized member of the Church and have a long record of abiding by its rules. She would wake up in the night with cold sweats and would cry, thinking she had doomed herself and her child to death because of her inability to conform to Church standards. This drove her to return to the Church. Rick did not approve of this step and the arguments began. After months of disagreement, Joscelyn placed Danielle in childcare, with their friend Maria, and went back to work.

Maria and Joscelyn had spent a lot of time together since both were stay-at-home mothers. Maria's daughter, Marisa, had been born ten weeks before Danielle. This made the transition a little easier. At least Danielle would be with someone she knew. Joscelyn had relished her time at home and didn't look forward to going back to work at the time.

The first day she dropped Danielle off at Maria's was one of the toughest days of her life. She felt she had failed and Danielle would be the one paying the price. As time went on, they both adjusted.

In January of 1989, the divorce was filed with no disputes. On May 1st Joscelyn moved into an apartment with Danielle, now 14 months old. She was on her own again but with a child this time. She felt scared and alone for the first time since leaving home years earlier. Joscelyn had no plans beyond keeping a roof over their heads and food on the table. She had lost her ability to make money, or so she thought. That year she made more money setting up real estate networks, than she had in any previous year.

In January of 1991, she resigned from Team Marketing Concepts, and started her own business, CRN, Creative Referral Networks.

Joscelyn had no qualms about starting her business. The training Mr. Wellmaker had given her stood her in good stead. She was ready for the leap. The fact that she would be pioneering a new concept, networking, as a single mother with no real savings to speak of and a two-year old to support, did not seem to faze her. I greatly admire her drive, vision, courage, and enthusiasm. Joscelyn could manifest anything she wanted. Joscelyn had a very strong personality and was used to going after what she wanted. She usually got it too.

Joscelyn, being a Capricorn with a Leo ascendant and a Scorpio moon, would fare well in business. She had what it took to make a success of anything she set her mind to do. Joscelyn paved the way for me by putting the company in a financial position that would provide me with income for two years after I walked in.

Her photographic memory was invaluable in a business that dealt so much with people. Clients used to comment on how easily she remembered dates, times, places, and phone numbers. She had no problem taking risks. Her attitude was, "If it doesn't work the way I think it will, I'll just fix it when I get to it." What a great way to think, yet it got her into many jams. But, like a cat, she always landed on her feet.

The Spiritual Path

Joscelyn began her spiritual path in 1986, at the age of thirty, just after she moved to Midtown. She began with the study of astrology. She set out to either prove or disprove astrology to herself. Joscelyn had been studying temperaments since 1982 and felt that if the temperament chart matched the astrology chart then there was a valid case for astrology, at least in her mind. What she discovered warmed her soul. Not only did the charts match, but the astrology chart also gave much more detail. So she continued on. She went for readings and continued to study astrology until she met Rick.

One of the readings was with a local well-known psychic named John Sandbach. He channeled Jane Roberts who had channeled Seth. He told Joscelyn that he saw a sarcophagus with the face of a queen on it. He also said that she would begin a search for the truth behind the religions and would end up finding her own truth. How right he was! Joscelyn continued her studies until just before the birth of Danielle in 1988.

In 1991, the spiritual work began again and this time in earnest. She picked up where she had left off in her study of astrology and continued her research. In January, 1992, she met the man who would pave the way for her return to that home out there in the stars—the one she knew existed, but could not find. That man was Chris Becker, her nemesis, lover, and friend.

2

Jelaila

By Jelaila

Chris was a wonderful, energetic, driven man with a razorsharp mind. He and Joscelyn began challenging each other the moment they met. When Joscelyn first met Chris, in January 1992, her feelings were ones of intense anger and deep sadness. She always wondered why she had experienced such powerful and disturbing feelings upon meeting him, rather than joyful ones. I would find out why in my first regression a year later.

Chris was married, even though in his mind it was only on paper. When Chris and Joscelyn met, he told her his marriage was over and that he was only staying to keep his house and, therefore, needed his wife's income since his new business was not profitable enough for him to take a salary. He told her he wanted to marry her and he would, as soon as his business became successful enough for her to quit work and stay home. Chris did not like Joscelyn's business because it involved working with a lot of men. Chris was extremely jealous and insisted Joscelyn call him each time she went somewhere.

In June, when Joscelyn became pregnant, she believed Chris would marry her and, together, they would raise their baby and her daughter, Danielle. It came as a great shock to her when he told her he didn't want the baby. All her dreams of a happily-ever-after life with him came crashing down around her. She thought of keeping the baby but felt she couldn't do it on her own, not with another child already. She felt she needed Chris's support, and life without him (he had threatened to leave her if she had the child), looked pretty grim. What a classic case of co-dependency. So, she chose to

15

have the abortion and end the relationship with Chris right after she left the clinic.

I, Jelaila, walked in on June 23,1992, during her abortion. Actually, the walk-in occurred in the recovery room. During that 45-minute stay she had dozed off. I don't remember all the thoughts I had when I awoke, but I do remember the feeling of being a little disoriented, with my emotions in a turmoil.

In the afternoon when Chris came to pick Joscelyn up, instead of ending the relationship right then as she had planned, I continued the relationship. It took another three years before I could finally put it to rest. I felt a strange mixture of anger, love, and compassion for a man so afraid that he would betray the one person he professed to love so much.

Preparing for the Walk-out

Just after Joscelyn met Chris, she and I began exchanging places during her afternoon periods of rest, which she needed because she was experiencing chronic fatigue. She would lie down on her bed and place her arms across her chest in mummy fashion and feel herself lift out of her body, a practice she had just recently begun. Each time I entered I would continue familiarizing myself with her body. Sometimes a Being would accompany me and stand at the foot of the bed. He would coach me on what was to be accomplished during that particular session.

Joscelyn began having thoughts about ascending during this time. She would tell Marla (her business partner at the time), as well as others, how much she wished she could leave the Earth plane and go home. It wouldn't be long before her wish was granted. When the time came for the exchange, it was very smooth.

3

Walk-ins

There have been many walk-ins on Earth and there are many here now. You may be one of them. Walk-ins come into adult bodies most of the time, yet, sometimes they will walk into a child's body. The exchange is done by prior contract in almost all cases, including Joscelyn's.

Most walk-ins occur during a traumatic event in the first soul's life, such as an illness or accident. A few occur during sleep or during a bad headache. Can you imagine lying down because you have a headache and getting up a different person? Major personality change!

Walk-ins must wake up after they arrive. This can take anywhere from a few months to 20 or more years. Walk-ins are activated through codes stored in the body. In the case of a prior contract, the codes are in the physical body. In the case of non-contracted walk-ins, they are stored in the emotional body.

Walk-ins seem to find the relationships with family and friends of the original soul to be either enriching or challenging. Many walk-ins divorce within three years. One of the problems I had was feeling disconnected from Joscelyn's family. It was as if I didn't know them. I hear this is common for walk-ins.

There are different kinds of walk-ins. Some walk-in exchanges are between two very different people—the first soul may not have ever known the soul coming in. Then there are walk-ins where the two know each other and have a contract for the walk-in to take place. There are also walk-in situations where the two souls have done the exchange before, in other lifetimes.

That is the case with Joscelyn and me. We are like Siamese twins. I am the ninth-dimensional oversoul of Joscelyn. I walked into her body in order to carry out the assignment I was given by the Nibiruan Council. As in other walk-in exchanges, there are many differences between Joscelyn and me. Joscelyn was strong and dynamic; I am softer and more flexible.

Joscelyn spent most of her adult life as a single person with few friends. She didn't seem to have the time or the inclination for friendships. I, on the other hand, have many friends and treasure the time I have with them.

Joscelyn had trouble relaxing. I have no trouble there! Joscelyn needed to be the leader in all things, whereas I enjoy working in teams. Joscelyn could make split-second decisions but I have to think them through. I like wearing softer colors in my dress, where Joscelyn liked the bold, vivid colors. She liked suits with flair and high heels; I've sold them all! Joscelyn was a brunette and I am a blonde. This is just a sampling of the differences between us. Can you see how this would make you think you were going a little crazy?

Honoring and maintaining the physical body is another area of concern for walk-ins. Joscelyn was a brunette and very attractive, of medium build and slender. She was a striking person and very charismatic. The plain Jane that grew into the lovely lady. She left me a slim, trim, healthy body, in good shape. I have tried to maintain it as she would, yet I fell into the same trap that many walk-ins do. I abused the physical vehicle by eating many of the wrong foods and not exercising the way Joscelyn did. I acquired an addiction to cookie dough, among other things. I have since changed that by watching what I eat and listening to my body and its signals.

Once the time arrives for a walk-in to wake up, they have usually gone through a series of synchronistic events that lead them to the conclusion that they are a walk-in. At this point they begin the process of discovering their contract.

The Encoded Crystal Cluster

Joysia, Chief Genetics Engineer of the Galactic Federation's Sirian A Council, has told me that walk-ins are activated by a cluster of encoded crystals that are formed in the body while the previous soul is still incarnate. He said some clusters are larger and more prominent and are located in the vicinity of the body that have the most to do with the contract of the future soul, the walk-in. This is most of the time, not all of the time. Such was the case with me.

My cluster is located on the back of my right shoulder, just above the shoulder blade. The right shoulder concerns the male energy and the future. It has to do with the ability to move forward into the future without fear. As though on cue, I began moving forward very quickly when I walked in. I realize this was because I had recently arrived here and wasn't weighed down by the density and fear of Earth. Also, I have not lived many lifetimes here on Earth. I have been told by the Council that the majority of my lessons were learned on Nibiru and other planets. They say I come here only on assignments.

Those individuals who form crystal clusters for the future tenants of their bodies form them in different ways. Joscelyn began forming mine through extended periods of needlework. She took an interest in stitching in 1984, and it became a passion. I am told by Joysia that the cluster was fully developed and in place by 1987. Joscelyn spent hours stitching on 30-count linen. She would stop only when Danielle needed her, or her fingers froze, or her neck and shoulders, mainly her right, would burn badly enough for her to stop. Joysia said the codes for the crystal cluster were already in Joscelyn's body at the time of birth.

After the walk-in, the crystals in the cluster were activated at the appropriate time, October 27, 1996. I must say, though, this cluster caused Joscelyn and I, both, much physical discomfort. When that shoulder muscle would spasm, for whatever reason, it would cause intense three-day headaches on the right side of the head. Only chiropractic adjustments gave any kind of relief.

Bobbie, my massage therapist, also tried to release the frozen muscle, through deep tissue massage. She and I would retreat to her massage table once a week, where she would almost work up a sweat and I would yell and moan in pain in between our laughter. She made it easier to handle because we would chatter on about the things going on in our lives at the time.

After about five months we gave up. The lump wouldn't budge. Of course we didn't know then that it was a cluster that held my Jelaila codes. It wasn't until shortly after I awakened that we found this out. It happened through a good friend of hers, Kim.

Kim lives in Colorado and works quite a bit with walk-ins. She was in town one weekend, and I made an appointment to see her. She looked at my body, eyes, and matrix and confirmed I was a walk-in, but a different kind. She said it was as if we were Siamese twins. When Kim looked at my shoulder, she said I had a frozen wing, and the reason for the frozen wing was that my codes were stored in the cluster on my wing, thus freezing it in place. This knowledge came as a great relief even though I didn't know exactly what she meant by frozen wing. From that point on, the spasms became less frequent and the headaches associated with them less severe. Later, I would find out how to release myself from the pain entirely.

I hope this information about the crystal clusters can help some of you walk-ins reading this book. Just knowing what you're dealing with can lessen the pain dramatically.

The Contract

Walk-ins agree to complete the first person's soul contract before they begin doing the work for which they came. This was the case with Joscelyn and me. But, understand that all this was done without my knowing my true identity. I thought I was still Joscelyn, yet I was acting from both her tapes and my own. This is the reason for the confusion many walk-ins have. For example, I would make a decision, then change it. This used to drive Chris crazy. Joscelyn made decisions and stuck with them. Get the picture? Poor man! Imagine what he felt like, being in a relationship with a person who was functioning with two sets of tapes.

My contract with Joscelyn stated that I would:

1. Transition Joscelyn's company to one that is more spiritually based.

2. Transition Joscelyn's relationship with her family to one that was more loving and accepting.

3. Transition Joscelyn's relationship with her daughter Danielle to one that has less conflict and more compassion. I was to also help Danielle mature and to, eventually, give residential custody of her to her father.

4. Transition and complete Joscelyn's relationship with Chris. This would prove to be the most difficult of all because it would have to be done without anger.

We are the Nibiruans

4

Chris

What a relationship to walk in to! Over the next three years it would twist and tear my heart to shreds and confront me at every turn to take back my power. I would grow stronger and eventually reach the point where I could maintain my boundaries, hold my energy, and lead a worldwide mission. The relationship with Chris would also prepare me for the relationship I had always wanted, the one with my lifemate.

Joscelyn was a match for Chris when it came to power, but I was not. We would be able to be in the relationship for about two weeks at a time and then we would fight, split up, and not talk for anywhere from a week to three months.

When Joscelyn and Chris were together they didn't break up. In many ways they were like two peas in a pod. Neither one could make a true commitment. They both entered relationships with people who were either emotionally or physically unavailable. I remember in September of 1993 when Chris finally divorced his wife how it scared me to death! He came to me and asked me for a commitment that I felt, deep down, would be wrong for me to give. My attitude from that point was to enjoy it while it lasted. I loved him so much, but somehow I knew I could never marry him.

Chris was invaluable when it came to business. I was not the entrepreneur Joscelyn was and he helped me navigate the rough waters of the business world. He taught me how to set boundaries (setting limits with self and others) with clients and vendors and not back down. He taught me how to negotiate contracts in a win/win manner. He taught me a lot. I don't know what I would have done had he not been there. He was my mentor and emotional support, my sounding board and my biggest fan. These were just some of the reasons I loved him so much.

Chris also taught me a lot about owning my own power. I realize now that this was part of Chris's contract—to play the dark side for me. If I had not had someone to struggle against for my power, I would not have been able to reclaim it. For this I thank him, as well. It enabled me to integrate the light and the dark within me.

Chris held his power when it came to others. His weakness was his powerlessness against his own fears, which were many. Yet, the more I began to set boundaries that were healthy for me, the more we argued. I see now that in my learning and setting boundaries for myself, Chris was losing control of me. This frightened him and confused me, because he would teach me to set a boundary with others outside of the relationship, but not set the same boundary in our relationship.

A woman having strong healthy boundaries in a relationship was something he was not used to. He did not know how to play by those rules. My setting boundaries with him would be the demise of our relationship.

The final curtain call for the relationship came in June of 1995 when I set a boundary he couldn't live with, regarding Danielle—he wanted to remove her father from her life by having sole custody of her. I did not agree, not because I felt Chris wouldn't be a good step-father, but because Rick was and still is a good father. Also, losing her father would be a major emotional blow for Danielle. Once Chris realized I wasn't going to change my mind, he left. I want to add that Chris had some valid reasons for wanting to remove the influence of Rick and his family from Danielle's life, but, for Danielle's sake, I will not disclose them here.

The evening he left I got down on my knees and thanked my guides, but it was still very painful because I loved him and still do. No matter how much I try, I cannot stop the feelings I have for him. I guess this is what is meant by the saying "true love never dies." Many was the time we would sit together and cry because we couldn't figure out how to make the relationship work.

Recovery from that relationship was slow, but I finally was able to fulfill that part of the contract when I realized he was being the only way he knew. I saw first hand how fear can create so much pain between two people in love.

Later I would come to understand why Chris had come into my life. I also understood what a great sacrifice he had chosen to make. I found out through Devin, that Chris, as one of the 90 Game Engineers[1], had chosen to play the dark-side role in order for me to be able to walk in, to learn to hold my power, and to wake up and fulfill my mission. Chris was my lesson mate.

Without him I might not have done it. He was the tumbler of sand and I was the rock he polished. By being with him I was able to sand off the rough edges and hone my power to the point that I could be a leader. The battles and power struggles were all worth it. He used to say that deep down, he believed he would never have me in this lifetime. He was right. His losing me was to be his sacrifice for me. I will be eternally grateful. As far as I'm concerned, the karma of our past lifetimes is finally cleared by this one great act of love.

Danielle

The relationship with Joscelyn's daughter, Danielle, was easy to change, because Danielle and I are "twin flames." Danielle was a cute, out-of-control four-year-old when I arrived. She had light brown hair and big brown eyes, and she was bright and manipulative. Danielle, born on March 15, 1988, was known as the "Miracle Baby of Olathe Medical Center." She was given this nickname because she had been oxygen-deprived for the seven minutes just before delivery, yet had been born perfect. The doctors believed there was a good chance she would be brain-damaged from the prolonged lack of oxygen.

Joscelyn had been in and out of labor for four days and still had dilated only one centimeter. She was also three weeks overdue. The

---

1. The 90 Felines and Carians who came to our universe to set up and manage the chosen Univeersal Game of Polarity Integration.

morning of Danielle's birth, Joscelyn had begun labor again, just after Rick had left to pick up her mother at the airport. His nerves were frazzeled from the repeated bouts of labor so he had decided to call in "the mother-in-law" to watch Joscelyn, so he could go back to taking care of the business.

No sooner had Rick gotten on the freeway when Joscelyn began having a brown discharge, which meant that the baby was in distress. She called the doctor and he told her to come to the hospital right away. Joscelyn called her father-in-law and he came over and drove her to the hospital.

Joscelyn was admitted around 9:00 a.m. and underwent a series of exams and stress tests by the nurses. At 1:00 p.m. she stopped laboring. At 1:30 p.m., Dr. Eidt tried to induce labor again, by breaking her water, but found he couldn't. The attempted induction managed to jump-start her labor. Dr. Eidt decided to let her labor awhile, to see what would happen. As a precaution, he had her prepped for a C-section. It was a good decision. Danielle had stopped breathing after the failed induction.

It seemed like the labor room was filled with doctors and nurses immediately. They quickly turned Joscelyn over on her hands and knees and began giving her oxygen. Within two minutes after the fetal monitor stopped, she was flying down the hall on a gurney, to an operating room. She was holding an oxygen mask to her face with one hand and pulling on the hospital gown to cover her behind with the other hand!

The doctors quickly knocked her out with gas and started to operate. They didn't even wait to complete the normal pre-operative procedures. Rick was in the operating room with Joscelyn, having been prepped with gown and mask just after the attempted induction. He noticed the veil over the baby's face which the doctors had to remove to scoop her out. Once removed from the womb, Danielle was whisked to the nursery to pump the meconium from her lungs and to check for brain damage.

After two Apgar tests she was declared normal! Dr. Eidt and the maternity ward staff were astonished. Word spread quickly through the hospital and doctors and nurses came daily to the nursery to examine the baby. Everyone wanted to see the baby who, by some miracle, had been born normal, after a harrowing emergency C-section birth. It seems Danielle had a reason for being born and a little oxygen deprivation wasn't going to stop her!

After the second day of examinations, Joscelyn was finally fed up with doctors poking and prodding Danielle, and demanded that the baby be left with her in her room. She also asked that Danielle sleep with her in her bed even though she was in pain and still on morphine. Joscelyn stopped taking the morphine that night, because she was afraid they would take Danielle back to the nursery while she slept. Although the maternity staff did all they could to put her mind at ease, she still didn't trust them and stayed awake most of the next three days, catching catnaps when she could.

On March 20th, Joscelyn and the Miracle Baby were discharged from the hospital. Joscelyn was more than ready to go home!

Later on, I was to find out that Joscelyn had acted as a sort of, surrogate mother. I understand now that Danielle could not have been born to me because my codes do not permit childbirth at this time. So, the only way for us to be together was for her to come through Joscelyn. I dearly thank Joscelyn for this.

Her relationship with Danielle was not the smoothest. They tended to butt heads a lot, but Joscelyn truly loved Danielle. Theirs was a battle of wills. They both had Leo rising. I find it interesting that Joscelyn always believed she would not be with Danielle for very long. She left that message in her memory tapes for me, another indication to me that I was a walk-in. If you are a walk-in, check your memory tapes, you may find similar messages.

Joscelyn spent much of her time working, since she had just started a business and was also working a second job to make ends meet. Danielle was only four at this time and spent most of her time with Grandma or Daddy. It got to the point that Danielle didn't want her Mommy to pick her up from daycare. She would cry and demand to have Grandma come pick her up from daycare. This is what I walked into.

I slowly began to mend this situation by giving up my second job and scaling back my hours. I felt that if I didn't, I would lose her.

This happened over a twelve-month period. When Danielle entered preschool, and then kindergarten, I would leave work early—not hard, since I had no burning desire to be there. I would pick her up, and we would go somewhere and play. This became a detriment in kindergarten because she didn't attend enough to pass. So, she had to repeat kindergarten because we played too much. Oh well, there could have been worse reasons.

One thing Joscelyn and I do have in common is our lack of respect for the educational system in this country. I felt I could teach Danielle more about life than she could ever learn in the classroom. That may be true, but it wasn't the curriculum that would pass her on to first grade.

During her second year in kindergarten I realized that the best thing I could do for her would be to give her father residential custody. Rick is very intelligent and school work is a breeze for him. He is also much more scheduled than I. I believe that children need a good foundation with some degree of consistency and scheduling. I realized I was not the one who would be able to give it to her. So, after much deliberation (working through my fears), I made the decision and signed the papers. A short time later, I was told by Devin, my guide, whose real identity I would later be given, that giving residential custody to Rick was part of my contract with Joscelyn. Knowing this made me feel a lot more comfortable with my decision.

Since then, our relationship has continued to improve. Danielle tells me how much she loves me and says she wouldn't want anyone else for her mom. There are no more power struggles. I'm teaching her the spiritual part of life and it has made a difference in the way she handles the things life puts in her path. I have learned to tell people, when they ask, that Rick gives her the foundation and I give her the wings to fly.

Joscelyn's Family

I transformed the relationship with Joscelyn's family to one of polite distance. In other words, we speak but don't go see each other very often. Joscelyn's family is scattered between Alabama, Tennessee,

Texas, and Colorado. That makes not spending much time together much easier. It's not that I don't like them, it's just that I don't have much in common with them. Most of her family still belong to the Church Joscelyn grew up in and can't relate to me at all. So, to spare them, I don't spend much time with them. I let them believe what they want about me. Joscelyn, being the black sheep, makes it a little easier. They expect her to do radical things.

Joscelyn warred with them constantly because she would not buy into their fear-based beliefs. She left the Church and moved away. Many was the time she tried to stay in the Church, even going so far as to get baptized just before her first marriage. But each time she eventually left. There wasn't any place for a single, attractive woman who ran her own show. This went against the Church's teaching that women were to be subservient to men and, most of all, were to be married, or in some way powerless. Joscelyn just couldn't go along with their program.

I am able to communicate with one of her brothers, Mike, because he seems more open than the others, and a lot more accepting. Mike and his wife, Debra, are also a lot of fun. I realize that Mike, and the rest of the family, are where they need to be for a reason, and I honor that. Yet, where they are and where I am are two different places. So I keep my distance and am polite when they call.

Creative Referral Networks (CRN)

The contract between Joscelyn and me stated that I would transition her business to make it more spiritual. After Chris, this was the next biggest challenge. It took me a little more than four years to accomplish. Devin and I, along with Marla, my new business partner, began steering the business onto a course of bringing spiritual principles to the business world through networking. I began taking interest in the company again. We taught business people a spiritually-based process of networking where they learned how to *give* first, in networking, instead of trying to *get* first.

We taught them how to build relationships based on trust and friendship before doing business with, or referring, each other. They learned to look within for answers, including how to handle relationships, and this brought many to an awareness of their Higher Selves and many began their Spiritual Path. Marla is carrying this mission forward and is now expanding into the corporate sector as well as working with singles groups.

The time came when I could begin integrating spiritual teachings into the training. From that point on, the business went through a series of changes that took Marla and me along with it. Today, it is a well-respected company with a promising future, and Marla is running it. She has become a true entrepreneur and has great vision. She also changed the name to Creative Referral Networks and Training. A good move in my estimation, as it more clearly describes what the company does.

Picking Up the Spiritual Path

I began researching astrology, just where Joscelyn had left off six years earlier. I bought so many books I felt as though I were keeping the local metaphysical bookstores in business. As soon as I had gleaned the required amount of knowledge on one subject I was impelled to go back to the bookstore and find the next book I was to read.

It felt as if someone were directing me; as if I was in school but my teachers were invisible. Never once did I take a course or class from anyone on anything. All my training was given to me by my guides, and there were a lot of them. Some guides would come for a short time—just long enough to teach me something—and then move on. It was like something was driving me forward to learn, to remember as much as I could, as fast as I could. I was moving at warp speed.

Nothing shocked or frightened me, except seeing entities, and that was from the old tapes left by Joscelyn, regarding demons— more Church indoctrination. Everything I read made perfect sense to me, and I began to think that maybe something was wrong with me, because I saw so much fear in people about the things I was researching and learning. They were afraid to look in these areas and to open their minds.

Anyway, I progressed from astrology to numerology, to soul mates, rocks, crystals, and spirit guides. From there I researched past lives and regressions. I found Edgar Cayce's and Shirley Maclaine's books at this point. Once I read those, I knew I had found my primary area of interest, though I looked at everything. I was also triggered by the Barbara Marciniak and Barbara Hand Clow books. They both channeled Pleiadian material, and something inside me told me that these were like messages from home for me. These and Sitchin's books did more for me, in terms of leading me to my awakening, than any of the other numerous books I had read.

The only thing that did not interest me was UFOs. I somehow knew they were real and the idea of extraterrestrials visiting our planet seemed so logical to me that I didn't feel the need to look into it. Being a walk-in, that makes perfect sense, since I am an extraterrestrial from the 9th dimension.

It was in early April of 1993 that I met my friend Bobbie, a psychic. I had gotten her name from Marla, who had gotten it from someone else. I called her, and we set up a time for a reading. It was mainly a status-check type of reading. Shortly after that, I called to set up a time for a past-life regression. I had just recently started reading about them.

And so began a series of regressions that would span the next twelve months. During that time I learned that Joscelyn and I had numerous lifetimes with Chris, from Atlantis, to Egypt, to England, to New Orleans, to Nazi Germany. Not one of them was easy and in the end we always parted on bad terms, usually with the early death of one of us. The lifetimes were filled with that same intense passion that carried over into this lifetime and is still there.

5

Devin's First Contact

Devin had begun communicating with me in April of 1993. It began
with a dream in which I was taken to this beautiful garden by two
angelic-type beings. Before this happened, a man pushed my tongue
aside and began speaking through my mouth. He continued
speaking, but I didn't care, because I was in this really cool place. I
didn't know what he was saying, either, but it didn't matter. At some
point he tried to give me his name.

When I awoke, the dream so disturbed me that I stayed awake for
hours. I almost thought I had been possessed by a demon. The next
day I told Marla about it and she suggested I again call Bobbie for a
reading. When we got together a few days later, I told her about the
dream and she said that this was a very powerful being, a master.
She also told me that a mutual friend of ours had a dream about him
and said she had picked up the name "David."

In any case, this dream stayed with me and now, three years later, I
realize it was a message for me about my contract to be a channel for
Devin and to assist mankind in learning about their history and the
DNA recoding process.

Devin and I communicated weekly from that time forward, and we
have evolved our relationship from my seeing him as some great,
perfect God-type being to seeing him as a friend and brother. We
argue sometimes, and I'm not always sure who wins. I'll tell you
more about Devin later, but let me say that there is a deep love and
respect between us.

First Channeling Experience at "Networking Kansas City-Style"

The first time Devin channeled through me before a crowd was at an event we held called Networking Kansas City Style. It was June of 1993. Marla and I had worked for months on this event. It was to be like a big debut for our company, Creative Referral Networks, Inc. We gathered corporate sponsors and bought banners and booked the hotel. We felt we were onto something big. Well, we were. People today still remember that event.

Our goal was to teach a large group of people one or two crucial networking skills and charge them for the event. This we did, and sold over a hundred tickets at $40 each. Not bad for a first-time event of that kind in Kansas City.

My role was to do the training that night, so I worked and worked on my notes and when the moment arrived, I stepped up to the podium and just went blank. I panicked, but then, all of a sudden, a calm came over me and I began speaking. An hour later people were coming up and saying how much they liked the event and congratulated me. I was speechless. Later, we went into the bar to have a drink to celebrate. That was when I realized that it was Devin, and not myself, who had done the training. Since then, he has taken the spotlight many times. People seem to enjoy his sense of humor and his witty remarks and analogies that make them think.

Little did I know then, that three years later I would again be channeling before large groups of people. This time, though, Devin would speak directly, and not only would Devin be there but also Joysia and a host of others from the Nibiruan Council. This time the subject wouldn't be networking, it would be DNA recoding and the history of mankind and Earth.

The Kryon Implant

In April of 1995, I discovered the Kryon books and the infamous Kryon implant, which is given to you in an energetic form. I asked for it right away. After I took the implant I was amazed. It was supposed to take 90 days to get through the period of darkness but

I was in it only three days! I asked for it on Sunday and on Tuesday morning I was awakened by a deep, male voice saying *Joscelyn!* It seemed to be coming out of my heart chakra. What an experience!

Kryon said in the book that after you receive the implant, you will be given three master guides. For months afterward I tried to get their names but, as much as I concentrated, I could not translate the sounds. So I called them Rayora, Amalee, and Devin. The first two turned out to be Rayshondra and Andralea. Rayshondra was Joscelyn's guide and maintained the body for me after I came in. She was in near-constant communication with Joscelyn regarding me and the body for about nine months after I walked in. Joscelyn was assisting me to seat into her body through Rayshondra. Seating into another person's body takes time. The soul of a baby has nine months to do the same thing.

There was a period of time when I had no communication with my guides after I walked in. This was for my sake, because hearing from them might have made me change my mind and leave Joscelyn's body. Rayshondra received information from Joscelyn about the existing codes and how to plug me, Jelaila, into them. To wake up then would have been detrimental to the mission. This was the reason for my extreme state of confusion for about a year. It was the reason I could not make decisions and stick with them. I was using her tapes and mine at the same time. What a mess!

Poor Rayshondra was working overtime. Rayshondra's hair is long and white. I often wonder if it turned white during the time she was taking care of me. My moving at such an accelerated pace didn't help matters any. I was like a child who went non-stop from the time it got up to the time it went to sleep at night. Had I slowed down a bit, it would have been easier for all concerned, but that's not the kind of person I am. This was a trait that Joscelyn and I had in common.

In May of 1995, after receiving the implant, I began to explore the history of Earth and the universe with much more fervor than anything else. It would consume me from that point up to the present.

I began reading Mark and Clare Prophet's books on the life of Christ and the Forbidden Mysteries of Enoch. The books that really turned the corner for me were the ones by Zecharia Sitchin. His books made me feel as though I had found home. They led me to the discovery of the Pleiadian planet, Nibiru, and its role in Earth's history. Sitchin's work also led me to the discovery of Devin's and my involvement with Nibiru and the creation of mankind.

6

Divine Reunion

In late November 1995, I finally made a complete contact with my guides Rayshondra, Grey Eagle, and a new guide, Andralea. It happened after I tried a technique Minta, my therapist, told me to use to find my quiet place within. I had begun therapy because of my increasing problem with making decisions and later changing them. I had not been able to find the quiet place up till then and, as a result, had trouble meditating for any length of time. I tried it in her office, and it worked. I couldn't wait to get home and go to that peaceful place again.

I got home, finished my chores, and settled into my favorite chair. I visualized myself floating over the Earth and when I landed, I was in the woods by a little creek. I was a little girl of about six, with blonde hair. I was playing in the creek picking up pebbles. I was watching the sunlight play over the water and reflect off the rocks underneath.

As I looked up, I saw a woman on my right. She was an older woman, a little on the plump side, with long, white, straight hair down to her waist. She had the kindest blue eyes I had ever seen and an even sweeter smile. She began to whisper her name in my ear. She did it by giving me the syllables and a picture to go with each. The first one was ray and she gave me the picture of a sun's ray. Then shan and she used a sheep. That was a hard one. Then dra and she showed me a drape. Finally I put it together. It was Rayshondra.

As I looked to my left I saw a tall, powerfully built Indian with long, straight, black hair. He had intense black eyes with thick, black brows shaped like hawk's wings. He, too, whispered his name in my ear. He started with the word grey and showed me a grey bird. Then he showed me an eagle. His name was easier to guess, though I was confused as to whether it was Grey or Great Eagle because the bird was so big.

Then I looked across the stream and saw Devin and Grandfather Two Moons. I already knew about Two Moons, as I called him, because I had had an artist friend of mine, Phyllis, in Sacramento, draw his likeness a few months earlier. Phyllis says she is able to draw the pictures by going into trance and allowing the people to come forward and present themselves as they wish to be drawn. She says they patiently pose until she is finished.

The man in the picture appeared to be Indian, because he also had long, black hair, high cheekbones, and brown skin. Only he was much shorter. I got the impression he was from a tribe that lived between the time of the Lemurians and the Picts. He said he was a spiritual leader of his tribe and that he had taught me, in that lifetime, about the stars and planets and the civilizations that lived on them. He said he was my grandfather and his name was one I would not be able to pronounce, so I should call him by any name I wished.

The night I received his picture from Phyllis, I stood looking out my bedroom window. It was a full moon but instead of seeing one moon, I saw two (an optical illusion), so I named him Grandfather Two Moons.

On Grandfather Two Moons' right was another woman, but this one was much younger and had long blond hair and striking blue eyes. She said her name was Andralea. This name, too, took a while to get.

On her right was a young Indian warrior. In his hand he carried a snake. He began whispering his name in my ear with a hissing sound. It did not take me long to get his name. I called him Snake Dancer. He was very handsome with long, black hair just past his shoulders. I would say he was about 5'10" and very strong and muscular. He also had very intense black eyes. Snake Dancer told me that he would be with me in the near future, in this dimension. He would be my future lifemate, an aspect of Enki, the creator and benefactor of mankind and son of Anu, ruler of Nibiru. Together, we would work to free mankind from fear by bringing them the knowledge and tools to regain their power.

It was a joyful reunion, and I cried and cried as I sat in my chair. They embraced me and told me to remain strong, that they would be with me. They were watching over me and told me to call on them at any time. They were just as anxious to reconnect with me as I was to be with them. The reunion lasted for about 45 minutes and then I dried my tears and went to bed. This happened just a few weeks before I awakened.

The Awakening

I woke up on December 1, 1995. It occurred as I was walking through the woods at the park near my home. This is a three-mile walking trail, and I had come out that day because I was being urged by Devin to go where he could talk with me more clearly. You see, the electricity running through our office with computers, fax, copiers, telephones and the like, makes it hard sometimes for people on the other side to communicate with us. Devin calls it having too much static on my channel.

I noticed that, once again, my neck was hurting. This had been a common occurrence since the walk-in, and it kept me from walking the trail as far as I would have liked to have gone. Eventually I had to turn around because my neck was beginning to hurt so much. In one moment of pain and sheer frustration I asked Devin why I had this pain and why it wouldn't go away. I had spent so much money, not to mention time, in doctors' offices with this condition, but nothing seemed to cure it. At that very moment, I even remember the tree I was passing at the time, I heard a very clear message from Rayshondra and Devin. They said, "It's because you are a walk-in, silly." Somehow that message did not surprise me. The rest of the way back to the car I was busily putting the clues from the past three and a half years together. From the trauma-filled operation when I walked in, to the gradual changing of personality, the inability to make decisions and stick to them, to the lack of enthusiasm for the business. All the pieces and clues pointing to the fact that I was a walk-in were there!

This momentous event occurred after much effort on the part of my guides. They must have been worn out by the time I finally got the message. One thing that seems to be fairly common with walk-ins is that they have neck problems. This is due to the fact that the new soul comes into the body through an opening in the neck.

An interesting side note: My business partner and friend, Marla, is also a walk-in, and I was the one to figure it out first. Here I was telling her at that time that she was a walk-in and I wasn't. Doesn't it just make you laugh?

I went home after the walk and took a bath. During this time I was in almost constant communication with the guides. They were giving me more information. Joscelyn was also there. Her presence was like a warm blanket wrapped around me. She had come to witness the awakening. It was a joyous occasion for those on the other side, since it meant we could finally put into practice the plan we had developed. The journey of enlightenment would be swift from that point on.

In early January 1996, I called Phyllis and requested a portrait of Devin. When I received it in the mail I couldn't wait to get the tube open. When I pulled it out and unrolled it, I nearly jumped back from the impact. His eyes sent an electrical shock right through me! He was older looking, handsome, and had a long beard and mustache. His hair was long and gray. The most powerful feature was his eyes. They were an intense blue and his expression left no doubt that he was accustomed to managing and directing people.

He chose to present himself that way, he said, because I still treated him as some highly revered father, so he went along with it. I finally took him off the pedestal, for which he was very grateful, and the next picture I had done of him was completely different. This time he showed himself as a young man with intense green eyes and long, dark brown hair. This represented another lifetime we had as lovers. I have no doubt I enjoyed it. He was a hunk!

I also had my lifetimes as Hatshepsut and Nefertiti drawn. In fact, I had found out about the Nefertiti lifetime first from having the picture done. I had asked for a picture of me in my lifetime as an Egyptian queen, the queen John Sandbach had seen on the sarcophagus years earlier.

When the portrait came, I was shocked—there before me, was a picture of Nefertiti! I called Phyllis to make sure that the picture was of the person I thought it was. She said she had seen two queens and two lifetimes but this one was the one that I was to receive at this time. Phyllis said when she came out of trance, she went to get a book on Egypt to see which queen she had drawn. She opened the book at random, and when she looked down, there in front of her was a picture of Queen Nefertiti! I took the portrait to my friends and asked them who they thought it was. They all agreed it was Nefertiti. They said we had the same eyes. As you can guess, this was hard for my ego to accept. Later on, I had the other lifetime done, and it was Queen Hatshepsut. She wore a Pharaoh's headdress, with eye makeup signifying the Priests of Amon.

Within two weeks after awakening, I was in a real identity crises. Suddenly, the clothes I had been wearing didn't appeal to me any longer and I felt my hair color wasn't right. So I began a swift transformation. I changed my hair color to blonde; I went from wearing suits and heels to long, flowing dresses, skirts, and flats. My color scheme changed, and my favorite color became pink, a color Joscelyn loved but never wore. She used to say, if you can't wear it, decorate with it.

Finally, in late January, I calmed down after having settled into my new look. I felt more like me now. My contract with Joscelyn is complete, and now I am moving on. I have done all I agreed to do for her. I feel Joscelyn would be pleased.

Part Two

History of the
Universe and Earth

By Anu/Jelaila

7

Nibiruan Council Members

9th Dimensional Royal House of Avyon

The Royal House of Avyon, of the Amelius Line, is from the Lyran planet of Avyon, the original birthplace of the Humans. Sananda, one of the 12th dimensional Council of Nine, is the Founding Father of the Amelius Line. When the Council of Nine completed construction of our universe (with the assistance of Felines and Carians), they asked another group of Felines and Carians to come here to set up and oversee "The Game" for our universe. The game chosen was Polarity Integration. Ninety Felines and Carians arrived to carry out the job. The ninety birthed into the Amelius Line (Felines) and Lucifer Lines (Carians) on the 9th dimension. All of the ninety members later became known as the 9D Nibiruan Council when it was formed as part of the Galactic Federation. Council members below the 9th dimension are aspects of the original ninety founders and are individuals in their own right. For the sake of clarity and brevity, only those members that are spoken of in this book are mentioned below.

Devin - 9th dimensional Patriarch of the House of Avyon and brother/husband of Jelaila; oversoul of 6th dimensional aspect, Anu.

Jelaila - 9th dimensional Matriarch of the House of Avyon and sister/wife of Devin; oversoul of 6th dimensional aspect, Ninhursag and 3rd dimensional aspect, Joscelyn Kelley. Presently in the body (walk-in) of Joscelyn.

Satain - 9th dimensional son of Devin and Jelaila; oversoul of 4th dimensional aspect, Marduk, current ruler of Nibiru.

Kavantai - 9th dimensional brother of Devin and Jelaila and brother/husband of Shalandrai; oversoul of 4th dimensional aspect, Nabu, current second commander of Nibiru, son of Marduk, and 3rd dimensional aspect, Barrie Konicov.

Shalandrai - 9th dimensional sister of Devin, Jelaila, and Jehaila; sister/wife of Kavantai; oversoul of 6th dimensional aspect, Enlil, and 3rd dimensional aspect, Susanne Konicov.

6th Dimensional Royal House of Avyon

Anu - former ruler/commander of Nibiru; Head of the Galactic Federation's 6th dimensional Nibiruan Council; father of Ninhursag, Enki, and Enlil; grandfather of Marduk; 6th dimensional aspect of Devin.

Ninhursag - daughter of Anu by Feline/Carian wife Rayshondra; Chief Medical Officer of the Earth Mission; 6th dimensional aspect of Jelaila and co-creator of mankind, along with Enki.

Enlil - son of Anu by Anu's half-sister Antu; prince of Heaven (Nibiru) and Earth; half-brother of Enki and Ninhursag; heir-apparent to the throne of Nibiru and Earth before Marduk took it by force.

Enki - son of Anu by the Dragon Queen of Earth, Dramin; half-brother to Enlil and Ninhursag; 6th dimensional aspect of 9th dimensional Jehovah (Jehaia); brother/husband of Damkina (Ninki); co-creator of mankind with Ninhursag and known as Ptah, ruler of Egypt.

Ninurta - son of Enlil by half-sister, Ninhursag; and grandson of Anu.

Adad - son of Enlil by official spouse, Ninlil; and grandson of Anu.

Nannar - son of Enlil by official spouse, Ninlil; and grandson of Anu.

Inanna - daughter of Nannar by official spouse, Ningal; half-sister of Ereshkigal; and great-granddaughter of Anu.

Ningishzidda/Thoth - son of Enki by Ereshkigal; grandson of Anu; also known as Quetzalcoatl.

Ereshkigal - daughter of Nannar by Snake princess, Id; wife of Nergal, half-sister of Inanna.

Nergal - son of Enki; Lord of the Underworld (African Mines), and husband of Ereshkigal.

Dumuzi - son of Nannar and brother/husband of Inanna

4th Dimensional Royal House of Avyon

Marduk - son of Enki by dragon princess, Damkina (Ninki); 4th dimensional aspect of Satain; leader of the Dark forces; present ruler/commander of Nibiru and Earth.

Nabu - son of Marduk, presently on Nibiru; great- grandson of Anu; 4th dimensional aspect of Kavantai, brother of Devin.

8

Anu Speaks...Introduction

Greetings to you, beloved mankind of Earth. I am Anu of your parent race, the Pleaidians, and more specifically, the Nibiruans of the Pleaidian battlestar/planet, Nibiru. I am sharing the history of your universe and planet to begin opening you to the realization that you are not alone in the universe and, indeed, have a race of beings who love you and watch over you.

What I share with you is an overview, a sort of main story line. I have pieced it together from the information I guided Jelaila to research. In your Bible, which we, your parent race, left to you, there is a line that says prove all things and this line, most assuredly, pertains to your history that I will share. At one point in the story, I will have Jelaila give you titles and authors, but I urge you to do the research yourself. This is the only manner in which you can prove it to yourself.

I would like to begin by explaining a little about myself and my people, your ancestors.

I am known by the name Anu in many of your excellent books about the Sumerian, Egyptian, and Babylonian gods of your planetary history. I am a Pleaidian of pure Lyran descent—a member of the Royal House of Avyon. Royal simply means we uphold the agreement our forefather, Amelius, an aspect of Sananda, made to maintain a pure DNA strain for the Human prototype.

As a race, we are tall, usually between nine and eleven feet. We have golden blonde or platinum blonde hair, blue eyes and fair skin. I, personally, am 9 feet, 11 inches tall with platinum blonde hair and blue eyes. The Lyran humans all had the original platinum blonde hair, blue eyes and fair skin. The golden hues of our bodies and hair were introduced by our mating with the lion people, or as some call them, the Felines.

My ancestors came from a planet called Avyon, in the Vegan system of the Lyran constellation. It was on Avyon that the human species was originally seeded and evolved by the Felines on the orders of the Founders and the Universal Spiritual Hierarchy. This is also where Sananda, one of the nine Founders of our universe, fragmented himself to create Amelius. Amelius was the first soul in the first human on Avyon. His line became known as the Amelius line—the Royal House of Avyon.

The Royal House of Avyon moved and settled in the Pleaides many millions of years ago in your time. Compared to other star clusters in the Milky Way Galaxy, the Pleaides are some of the newest planets and stars. Our ancestors, led by the family patriarch, Devin, were given the Pleaides as their new home by the nine Founders of our universe, after their original planet, the Lyran Avyon, became uninhabitable.

We are an independent breed of people. This was not always so. Before the creation of Nibiru, we were a people that expressed only the feminine qualities. Since living on Nibiru, we are experiencing the masculine side, which gives us our independence. As a race, we are working toward finding the balance between the two, the integration point as you would say. I would like to give you now, information on Nibiru itself.

Nibiru is a beautiful, red, artificially-enhanced planet—a planet much like your Jupiter that began its evolution into a new sun but whose progress was suddenly halted by a cataclysmic event long, long ago. I'll discuss that event in the next book/manual. The gold in our forcefield gives it a magenta hue. We live inside the planet, not on the surface as you do. The outside of our planet is encased in a metal-like substance not found on your planet. The protective forcefields around the outside of our planet/ship give it the brightness many spoke of in your past civilizations, including the Egyptians. They called Nibiru "the bright star of the crossing. The rings around our planet are part of the propulsion system that moves us through space and also adds to the brightness of Nibiru.

Nibiru was created by the Galactic Federation as a peace-keeping battlestar/planet. Its purpose is to promote harmony among the many diverse civilizations on the planets in our galaxy. The Pleaides are the current home base of all human civilizations in our galaxy, having replaced the Vegan system as home base a long time ago.

Nibiru is about four times larger than Earth. It has room for many races and species of beings to co-habitate in relative harmony, generation after generation. There are beautiful lakes, seas, oceans, mountains, and valleys, just like on Earth. Trees and plants of every imaginable kind grow in profusion. Nibiru was created to resemble our original home planet, Avyon.

Avyon had two suns and a firmament which made it a tropical paradise. Even though the light inside our planet/ship is artificial, Nibiru is still a lush, green paradise. It has a simulated day and night, with a canopy of stars in the night sky. Most of the vegetation on Earth came from seedlings developed in our extensive laboratories and propagated on Nibiru. We also have cities and towns just like you.

Since we are a peace-keeping battlestar, we have vast areas for the maintenance and storage of defense ships and exploratory shuttle craft. The Starship Enterprise, from your television show, Star Trek, is very similar in mission and purpose to that of Nibiru.

On a spiritual level, Nibiru provided a way for us, the feminine-polarized Pleaidians, to experience negativity. By having to protect colonies, we came face-to-face with negativity and were, therefore, able to experience and understand its underlying roots of fear. As a race, we had become stagnant for lack of negativity. Negativity serves a very useful purpose in spiritual growth. Since we had no negativity on our planet, we had no growth. The solution to our problem was the creation of Nibiru.

Nibiru is magnificently equipped with the most advanced technology in the universe. I was told, as a child, of the great day it was put into operation. There was much fanfare, feasting, and celebration. Nibiru was more than a battlestar. It was, to us, the physical representation of our forward quest for spiritual growth. It was also our new home. The launching of Nibiru was a sight to behold.

My forefather, Niestda, was the first ruler/commander of Nibiru. Seventeen generations later, I was given command of Nibiru after Alalu, my half-brother, was asked to step down by the Galactic Federation's Nibiruan Council. He had taken command upon my father's death and served nobly. He was a good commander, but could no longer fulfill the needs of the people and the position at that time.

Alalu had lost his wife and daughter during a battle with the Reptiles. He felt he needed to get away. Alalu was sent to Earth to look for gold. He enjoyed that kind of travel and it helped him recover from the deaths of his wife and child. I know it has been written that I wrested command from him in a great battle, but this was not so. That was my grandson Marduk's doing. Marduk had all the written records changed after he became ruler/commander.

I was the ruler of Nibiru until Marduk took it by force in your year of 2,200 B.C. I began my rule long before coming to your planet 480,000 years ago. I now reside on a Pleaidian mother ship along with my sister/wife Antu, my daughter, Ninhursag, my sons Enlil and Enki, and a host of other family members. We are currently maintaining an orbit in the vicinity of your planet, Saturn, where we have extensive laboratory facilities.

There are many beings from Nibiru and other planets and galaxies aboard this Pleaidian mother ship, who are actively involved in the fulfillment of the Divine Plan for Earth and its people. Many of these are the other, or extraterrestrial, parents of star seeds on Earth. Also, there are representatives of the parent races of starseeds and walk-ins aboard ship to give assistance to their incarnate children on Earth.

It is very exciting to work with you through our people on Earth. They are known as Emissaries of the Galactic Federation's Nibiruan Council and also as Avyonians. There are many of these emissaries incarnate among you who are spreading the word about your parentage and the good news of the assistance we bring in the closing hours of your final drama in a 3rd dimensional reality.

Soon, you will finish this great drama and move to a 5th dimensional reality and rejoin us, for most of us exist in the 5th and higher dimensions. Nibiru is 4th dimensional at this time. We, on the Pleaidian mother ship, are 6th through 9th dimensional and incarnate at this time.

Fig. 1

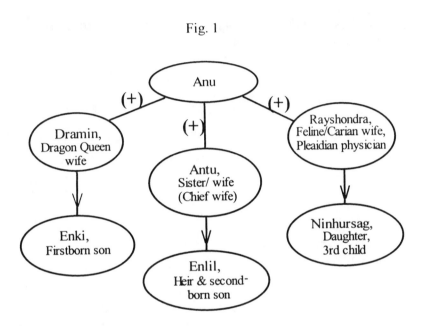

I head the 6th dimensional section of the Nibiruan Council of the Galactic Federation. Our main task at this time is to assist mankind in a process we call "DNA Recoding and Reconnection," though we are also working on providing you with the history of your universe and planet. We work with the Christos Office of Earth's Spiritual Hierarchy as well. I will provide more information on the latter in a moment. The Nibiruan Council is quite large and multi-layered and provides assistance to Earth and other planets in our solar system. We work with beings from many galaxies and star systems.

We are presently engaged in an alliance with our mentor race, the Felines of Sirius A, and the Christos Sirians. It is called the Sirian/Pleaidian Alliance. Together, we are working on DNA recoding. We also supply the information of your history as well as parenting you.

53

The Felines provide assistance by adjusting the DNA implants in your astral bodies and fusing the twelve DNA strands in your etheric bodies into your endocrine system. The Christos Sirians provide assistance in helping us reach those on Earth who are ready to embrace the recoding process. They work with the guides of those individuals, urging them to clear and cleanse in preparation for the process.

Now, I would like to take you on a journey back in time to share the history of your universe with you.

9

Anu on...Universal History

I will begin this story with a brief overview of how the universe is set up and managed.

Our universe is made up of many galaxies, stars, and planets. The Spiritual Hierarchies oversee all of this. There are many different hierarchies. The Universal Hierarchy is like the board of directors of a large corporation. The other hierarchies oversee the different galaxies. Still other hierarchies oversee the constellations and star systems, and beyond that are the hierarchies that oversee the individual planets and stars. These are the Planetary Hierarchies.

All hierarchies are made up of souls that have chosen to serve in the angelic realms. Amelius/Sananda, although not an angel, is in charge of your planetary Spiritual Hierarchy since he is head of the Etheric Sirians, the heirs of Earth. Another type of hierarchy oversees the Divine Plan of the different soul groups. I will discuss this next.

Souls

There are two kinds of souls, those that incarnate and those that don't. Those that don't incarnate are called angels. Those that do incarnate are called incarnates. All souls are divine fragments of Divine Creator, or as many of you say, God. Angels learn and evolve just like the rest of us, only they do it in a different way. They evolve through serving and ministering to the incarnates as members of the spiritual hierarchies. Incarnates evolve through learning to serve each other.

So, in essence, all souls evolve through service to one another. Service is synonymous with unconditional love and unconditional love for each other is what we all strive to achieve.

Soul Groups

Souls are divided into groups according to their preference for evolution (angel or incarnate) and their stage of evolution. Grouping incarnates makes it easier to manage them and their Divine Plans. Souls move in and out of groups as they evolve. Some souls evolve faster than other souls.

Soul Clusters

Soul clusters are clusters within the larger soul groups. Clusters consist of no more than fifteen souls who are together to work on a specific lesson such as impulsiveness or selfishness. Once the lesson is learned, the souls move on to other clusters to learn other lessons. Guides are attached to each soul cluster to assist them during their incarnations and later, when they return to spirit.

Earth's Soul Groups

At this time there is one very large soul group using Earth, the Earth Sirians. This particular soul group is the one for which the two-stranded DNA physical body was created. All souls on Earth at this time use the two-stranded DNA vehicles, regardless of planetary origination. There are many other smaller soul groups on Earth as well. The starseeds and walk-ins make up some of these smaller groups. Now let's look at the Great Divine Plan.

10

Anu on...The Great Divine Plan

Everything in our universe is a part of Divine Creator. Divine Creator, whom I will refer to as He for the sake of simplicity, was in a state of absolute primacy and perfection, and, after a while this got to be pretty boring. So Divine Creator decided He wanted to experience more of Himself. To do this, He had to fragment Himself into many pieces, and He did just that.

He fragmented Himself into multiple thousands of little Divine Creators. Each one was a carbon copy of Himself, and He called them First-Source Souls. Each had the ability to create, manifest, reason, and feel emotions of every kind. In essence, they were Gods just like Him.

This is why we tell you that you are Gods incarnate, just as we are Gods incarnate. Each of you is a piece of Divine Creator, just like myself and everyone else. Each of us is equal, since we have the same powers of reason, creation, and so forth. It is what we do with our power and how we feel about it that brings about the feelings of inequality.

These new First-Source Souls were divided into two paths of evolution, angelic and incarnate. The Incarnates have twelve dimensions through which to evolve and the Angelics have seven realms for their evolutionary process.

The angels and the incarnates could not evolve without each other, which meant they would have to work together for the evolution of all.

Fig. 2

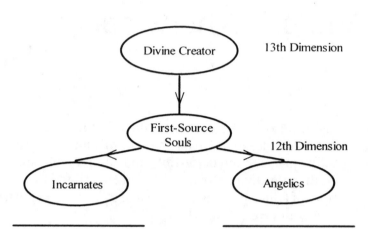

Our Universe

Incarnate Dimensions		Angelic Realms
		(Assist game overseers and players)
Game overseers	- 12th, 11th, 10th	
Game Players	- 9th	
?	- 8th	7. Supra Universal Hierarchies
?	- 7th	6. Universal Hierarchies
Felines/Carians	- 6th	5. Galactic Hierarchies
Pleaidians	- 5th	4. Constellation Hierarchies
Astral	- 4th	3. Star Systems Hierarchies
Earth	- 3rd	2. Solar Hierarchies
Animals	- 2nd	1. Planetary Hierarchies
Plants & Minerals	- 1st	

To evolve, the souls needed to experience something that would create growth; so Divine Creator created Games with the main Game being the one called Polarity Integration. The Game required the creation of Light and Dark roles. The task was to experience all facets of each and learn to integrate both—Polarity Integration. Once a soul achieved this, that soul would be reunited with Divine Creator.

Divine Creator also included a formula or tool to achieve polarity integration. That formula is called the 13th-dimensional Formula of Compassion. Since Divine Creator is the essence of love, the task was to learn unconditional love and compassion for all beings, regardless of the role they are currently playing. This formula would be given to all souls just prior to the final acts of their planetary Game. The Formula would be available to all souls on the planet. It would be used to permanently release the negative emotions from their bodies thus lightening them. Once enough souls completed this task, the planet would shift to the next dimension, thus shifting all the other planets ahead and behind them one step closer to reunion with Divine Creator.

Next came the creation of the Divine Plans. Each universe, galaxy, star, planet, and soul had a Divine Plan, and the angels in the hierarchies oversaw them all. The angels also had their divine plans. You have an individual Divine Plan for evolution. You are also working on the Divine Plan of your soul group, your planet, your galaxy, and your universe at the same time. No wonder you are so busy!

Our Universal Divine Plan

Our Universal Divine Plan was created by the Founders in conjunction with the newly organized Universal Spiritual Hierarchy and the Game Engineers, the ninety Felines and Carians. The Founders were a small group of First-Source Souls who had banded together. They are called, by many on your planet, the Council of Nine.

The Founders chose Divine Creator's Polarity Integration Game as the game for their soon-to-be-created universe. The Founders asked the assistance of beings from another universe who had already completed the same game. These were the Felines and Carians.

In their universe, the Felines had represented the Light and the Carians had represented the Dark. They were asked to construct the universe, create the lifeforms, including the physical vehicles for the souls, and seed the planets and stars. The stargates, dimensions, portals and grids would need to be created as well. The Feline

Universal Construction Engineers created the planets and the
Feline Genetic Engineers created the lifeforms, while the Carian
Magnetic Engineers took care of the stargates, dimensions, portals,
and grids.

Fig. 3

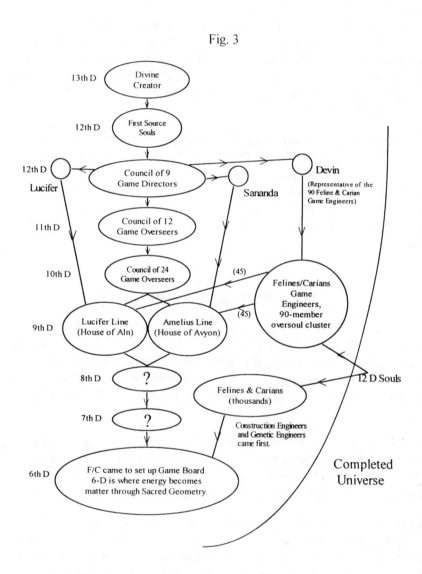

The two main races created for the Game were the Humans and the Reptiles. The Humans were created in the image of the Felines and the Reptiles were created in the image of the Carians.

After construction was complete, the Founders asked another group of Felines and Carians to assist. This time it was to set up the Game itself. These were the Game Engineers, and there were 90 that responded to the request.

All were seasoned players of the Polarity Integration Game, having taken uncountable numbers of souls in many universes, galaxies and planets through the Game. They were all First-Source Souls.

The Founders arranged meetings with the 90 Game Engineers and the plan for the Game was developed. It was decided that the 90 would split up and 45 would birth into the Lucifer Line of the Reptiles and 45 would birth into the Amelius Line of the Humans, on the 9th dimension, to begin the Game. This would input into the two races a genetic memory of a completed Polarity Integration Game. The Game Overseers would reside in the 10th and 11th dimensions as the Council of Twenty-four and the Council of Twelve. The Game Directors would reside on the 12th dimension as the Council of Nine.

Once everyone was briefed in their roles, the 90 Game Engineers chose one of their own to represent them on the Council of Nine. The one chosen was Devin. Devin is the ninth member of the Council of Nine. Devin's role was to begin the Game by birthing into the Amelius Line and becoming the Patriarch. After that his main duty would be to remain on the Council of Nine and awaken the remaining 89 at prearranged times during their completion of the planetary and galactic Games.

Once all the souls in the universe achieved polarity integration, the game would be finished and our universe would reunite with Divine Creator. This is what you and I are working on now. The completion of the Polarity Integration Game on Earth will move the universe one step closer to reunion with Divine Creator. I will now briefly explain the portion of the Divine plan of our Galaxy that pertains to you and me.

Our Galactic Divine Plan

Within the plan of our galaxy, there were to be four major races of beings involved in our Polarity Integration Game. I refer now to those who play a major role in this story. They are the Humans, the Reptiles, and their creators, the Felines and Carians. Although there are many other species of incarnate beings in our universe, they, like the Humans and Reptiles, are also offspring of the Felines and Carians.

The Divine Plan called for the Reptiles to represent the Dark Forces and the Humans to represent the Forces of Light as a whole, though each of us experiences lifetimes in both, the Light and the Dark, at some time in our evolution.

The Felines would create the original DNA blueprints for the Humans and Reptiles. The Carians would provide protection for each race until it could fend for itself. Each group of souls would learn from the role they played in the plan. It is much more complicated than this, but I will save the details for another book. This is merely an overview.

The Reptiles and the Humans would learn to move beyond fear, hate, and prejudice, and learn to activate the 13th-dimensional Formula of Compassion within their codes. Thus they would value love and appreciate each other's differences. This was and still is their evolutionary goal, and our galaxy is providing the stage to act out this drama.

Earth's Divine Plan

Once again, we have a multifaceted and complex plan. It includes the completions of the Avyonian Divine Plan and the Etheric Sirians' Divine Plan. The Etheric Sirians are the Lyran Humans who were moved to Sirius B and, eventually, given Earth as their home by the Founders. The completion of the Avyonian Divine Plan would enable the Etheric Sirians to complete theirs.

Before I proceed, it is necessary to share a little background on each race.

Felines

The Felines are the ones many of you call the Lion people. They are from a universe that has already completed itself. They arrived in our galaxy when it was first being constructed. They were invited by the Founders and the Universal Spiritual Hierarchy to be the master geneticists for our universe. They reside on a star in the Sirius constellation known as Sirius A.

Fig. 4.

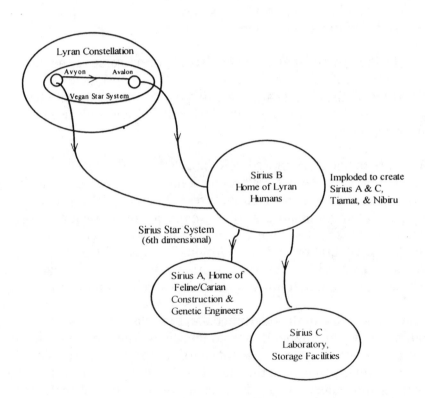

They are not to be confused with the Humans who colonized Sirius B. They were living on Sirius A long before the Humans from Lyra arrived.

I would like to interject a little history on the Sirius star system. There are three stars in the Sirius Star system. Sirius A, Sirius B, and Sirius C. Sirius B was the first star and when it imploded it created Sirius A and C.

When the humans first arrived on Sirius B from their sojourn in the human colonies of the Vegan system of the Lyran constellation, they were given much help by the Felines. Sirius C is currently used for storing and warehousing materials and supplies.

Tiamat was also created from the implosion of Sirius B, as well as Nibiru. When it was decided to create a battlestar, this huge chunk of rock floating in space near the Pleiades was located by my ancestors and turned into a battlestar/planet. So you could say that Nibiru is Sirian and Pleiadian.

The Felines are the master geneticists of our universe as they were of their own. They created the original DNA blueprints for all lifeforms on the planets and stars in our universe. This includes vegetation, animals, and the physical vehicles that souls use for incarnation.

They are tall beings with red-gold hair and usually, hazel or green-gold eyes. Felines are powerful, highly evolved beings with loving, gentle, and benevolent natures. When we get hurt in battle, we go to them for healing, as they are the doctors of the universe. They never take sides in any war, as they have already completed the lessons of integration and unconditional love. They also take care of souls who have suffered severe emotional damage in incarnations. These souls are sent to the Felines to be cared for after death.

The Felines exist in the 6th dimension to perform their work, although they are higher dimensional beings. The 6th dimension is the dimension of creation, where energy becomes physical. The Felines work closely with the Spiritual Hierarchies of the different planets and galaxies in our universe. When it is decided that a soul group needs a planet and a physical vehicle for incarnation, it is the Felines who are assigned the task of preparing the planet and creating the bodies. There have been times when they also created the planet, as well.

The Felines are heavily involved in your history. The sphinxes found around the world are a reminder of their existence and their involvement. Have you ever wondered why the lion is the king of the jungle even though he is not the largest? The Felines left the lions, and all cats, to be transmitters of information back to Sirius A. The lions, in particular, were made king of the jungle so that they would not be killed off and could, therefore, continue being transmitters of information throughout the many thousands of years that Earth and mankind would need to complete their Divine Plan. Cats are the communication link between the 3rd and 6th dimension.

Many of you are aware that often kings and queens had a cat of some kind for a constant companion. In ancient times the Pharaohs of Egypt were guided by the Felines through their cats. If you see someone who has what you call cat eyes, this is probably because they are partially of Feline origin.

The Carians

The Carians are a tall group of beings with birdlike features. They came from the same universe as the Felines. They lived on Sirius A along with the Felines. The purpose of the Carians is to act as a protector race for evolving physical vehicles on a given planet or star. It was the Carians who protected the evolving Lyran human primates from the Reptiles who wished to destroy them.

The Carians are also responsible for the creation of grid systems, stargates, and portals. They are the magnetic engineers while the Felines are the genetic engineers. The Carians also left a trademark of their involvement with us in the form of the Phoenix and the Eagle as well as the Raven and the Hawk. These birds were (and still are) sacred to many people.

The Carians work hand in hand with the Felines. They provide protection for planets and stars that are still evolving a land guardian race of beings such as Humans. Their task is complete when the evolving land guardians can protect their planet themselves. The Carians shouldered the responsibility for protecting the Human colonies in our galaxy before Nibiru was created.

The Carians do not use military weapons for protection. Instead, they use energy to create doorways and locks on portals going in and out of a planet, and stargates in and out of dimensions.

The Reptiles

The Reptiles are known by many as the Reptoids. They were created long before the Humans, on the planet of Aln, in the Orion constellation. The Reptiles were created first so that they could achieve technological superiority. They had already attained space travel when we, Humans, were just emerging from the oceans of Vega.

The Reptiles, also known as the Snakes, the Dragons, and the Lizards, (affectionately called by some, the Lizzies), come in a variety of styles, colors, and sizes. They are usually green, brown, bronze, black, or a combination of any or all of these colors. Their eyes are either green or red. Their skin is scaly or smooth and cold to the touch. They have the feelings of the five senses, but not all the emotions.

The home base of the Reptiles is the constellation of Orion and its neighbor, Sigma Draconi, where the Dragon people live. The Reptiles colonized many planets and star systems in our galaxy. They were given a creation myth that stated that they owned the galaxy and had the right to colonize any planet or star they desired. And if there was a non-reptilian race already present, they could and should destroy it.

Now you can imagine how the other races felt! This creation myth caused the death and destruction of many planets and people on both sides. But, let me remind you, the Reptiles were created by orders from the Founders to represent the masculine, or dark side, and the Humans were created to represent the feminine, or light side, in this Polarity Integration Game. *So please, do not judge them, for they are only performing their roles.*

The Reptiles have colonized Earth more than once. They initially colonized it when it was still Tiamat, and then again, after Tiamat had been split in half and became known as Earth. But, the second time was not a real colonization.

Instead, it was the reorganization of the remnant of Reptiles who had gone underground during and after the destruction of Tiamat. Your modern-day snakes and lizards are the descendants of the Reptiles. The dragons were also here and your myths of flying dragons were actually not myths at all. They were real!

The Reptiles are more technologically (masculine, dark) advanced than they are spiritually (feminine, light) advanced. This is what they are learning to balance, just as we Humans are learning to balance the other way.

The Humans

The Human race evolved in the Vegan star system of the Lyran constellation, on the planet Avyon. As I mentioned earlier, the Humans are created in the image of the Felines. They were given a creation myth different from the myth given to the Reptiles.

The Human creation myth specifies that the Humans are also to colonize any planet or star they choose but, if they find another race on the planet, they must negotiate a peace treaty and strive to live in harmony with the neighboring race.

Both Reptiles and Humans maintain a pure DNA strain for future seedings. These are the Royal Houses. The Royal House of the Reptiles is the House of Aln. The Royal House of the Humans is the House of Avyon, mentioned earlier.

Now that you have a little background on the four major races, I will continue my story. Let us begin at the point where the Founders were ready for the Felines to create the human species. Keep in mind that the Felines had already created the Reptiles and they had reached the stage of space travel by this time.

The Felines commenced seeding Avyon for the new race of Humans. The Humans began in the oceans and would eventually move onto land. Your Darwin was correct when he stated that man began in the oceans. At the stage of aquatic primate, the Felines took the majority of the primates out of the ocean and upgraded them to the stage of a biped human. The remaining aquatic primates, whales and dolphins, were left in the ocean to maintain the biosphere of Avyon. Biosphere maintenance is critical to any life- sustaining planet.

Once the humans on Avyon had evolved to the point of space travel, they colonized another planet in the Lyran constellation. They named the planet Avalon. The new colony was established as a feminine-polarized society in keeping with the societal system on Avyon. Soon the Reptiles arrived, bringing with them their superior technology and their masculine-polarized form of society. This was good because it created the first opportunity for polarity integration. Unfortunately, without sufficient spiritual knowledge and experience, dissension and conflict resulted. Allow me to explain.

The Reptiles felt threatened when they realized the Humans were moving out beyond Avyon. What if the Humans took over the whole galaxy? Where would they, the Reptiles live? In their minds this pioneering effort had to be stopped. At first, there wasn't a direct confrontation between the Reptiles and the Humans. That wasn't the Reptilian way.

Instead, the Reptiles did their usual thing. They began infiltrating the colony and sowing seeds of discord among the people, while, at the same time, courting the friendship and trust of the Humans by offering them their technology. This created a division between those who wanted to grow spiritually and those who wanted to grow technologically. Discord continued to escalate to the point of civil war, at which time the Reptiles jumped in on the side of the masculine-inclined colonists, and the result was the near destruction of the colony and the planet.

If you look at your history books, you will find evidence of Reptilian influence and tactics in your planet's wars and disputes first, sowing seeds of dissension and then taking sides, which leads to destruction.

After the near destruction of the Avalon colony, the Founders decided to move the Humans to Sirius B to continue working on the integration of the masculine/feminine polarities, but this time without interference from the Reptiles. This plan was a partial success.

What occurred was a greater division among the, now, Sirian Humans (their name was changed from Lyran Humans to Sirian Humans). A non-physical (etheric) human group, the Etheric Sirians, developed, who devoted themselves to the pursuit of spiritual knowledge and healing, accentuating the feminine polarity and Lyran way of life. The leader of this group was none other than Amelius. The rest of the Sirian Humans chose to remain in a physical state. Those in the physical state were the masculine-polarized humans called Physical Sirians.

Once again the Founders and the Spiritual Hierarchies felt it necessary to move both groups of Humans to new locations to continue working on polarity integration. This time, the chosen planets were Aln in the Orion Constellation, and Tiamat.

Since the Physical Sirians were masculine-polarized, they were sent to Aln, home of the Reptiles, to establish a new colony, practically in the Reptilians' back yard. This group became the Orion Humans. It was hoped this move would enable them to better understand themselves and their masculine polarity by being on the same planet with the Reptiles. But, this did not make the Reptiles very happy, and it wasn't long before the battles began and the Alnian colony was nearly destroyed. The Human colonists were forced into slavery by the Reptiles.

Eventually, the Alnian colony was rebuilt and the drama of polarity integration continued, but now there was a new player in the game—the Black League. The Black League began as a small band of Orion Humans who formed a resistance movement to overpower the Reptiles who had enslaved them. Later, a group from the Black League would escape from Orion and travel to Tiamat, looking for a fresh start.

The Etheric Sirians from Sirius B, led by Amelius, were given Tiamat as their new home. The Etheric Sirians would eventually need to become physical again to perform the duties of a land-guardian race for the planet. So the Felines seeded the planet to fulfill that need.

When the Etheric Sirians arrived, they found the new land guardian race at the aquatic primate stage. The Etheric Sirians took on the role of guardian for this new race and lovingly watched over their future physical vehicles, tending to their spiritual and physical needs.

Life went along as planned until a group of the Etheric Sirians began spending time in the bodies of animals on the planet. After a while, this became a real concern to the remaining Etheric Sirians. They were well aware that thought becomes reality and if their brothers and sisters continued to spend time in those bodies, they would forget they were etheric and become stuck. Mankind on Earth has experienced this same situation. Mankind has forgotten they are souls in physical bodies. Instead, they believe they are physical bodies with souls. This occurred with the Etheric Sirians.

Fig. 5

Human Lineage

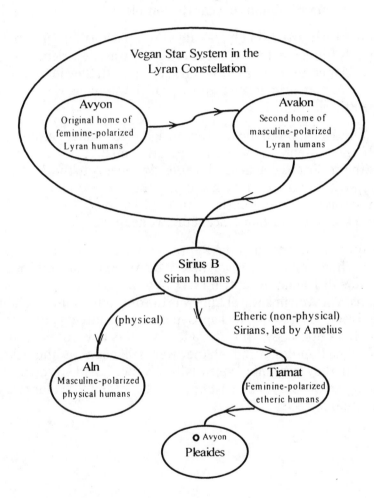

Due to the Cosmic Law of Free Will, the remainder of the Etheric Sirians could not stop their brothers and sisters. So, a plan was made to rectify the situation. A portion of the Etheric Sirians would become the Christos Sirians and form the Office of the Christos. It would be attached to the planetary Spiritual Hierarchy and would oversee the rescue operation for their brothers and sisters in the animal bodies (Earth Sirians). This plan, like most galactic plans, would take many millions of years to complete. Back to the Reptiles.

The Reptiles heard through the galactic grapevine that a new planet was ready to sustain life. As they believed it was their right, they set out for Tiamat to colonize it. When they arrived, they found the evolving human primate civilization, which was being watched over by the Etheric Sirians.

The Founders allowed the Reptiles to colonize Tiamat because they felt this new attempt at polarity integration might prove successful. The Etheric Sirians began sending positive energy to the Reptiles in the hopes it would soften them and help them let go of their creation myth. If this were accomplished, the Reptiles could work and live in harmony with the primate race. For awhile, it did.

Shortly after the arrival of the Reptiles, Amelius asked Devin to leave his home on Lyran Avyon (See "The Amelius Line" on page 72) and come to Tiamat to contribute DNA to upgrade the primates to humans. The Amelius Royal Line was the only pure human strain in the universe, so it had to be kept separate to ensure its purity for future DNA upgrades. This upgrade made the, now Humans, more equal to the Reptiles. Although they were still not as technologically advanced, they were more spiritually advanced, and spiritual maturity on the part of at least one race is needed for polarity integration to occur.

11

The Amelius Line

The Royal House of Avyon

Amelius was the first soul to incarnate into the body of the first physical human vehicle on Avyon. Amelius established a line on Avyon, as his later descendants, Devin and Abraham, established lines on Pleiadian Avyon and Earth respectively. Amelius upheld the initial agreement made between his oversoul Sananda and the other eight Founders. He maintained a pure lineage and DNA for the human prototype created in the image of the Felines—tall, platinum blond hair, blue eyes, and white skin.

After Devin and his family completed their assignment on Tiamat, they moved to the Pleiades to establish the Amelius line there. They chose a planet for colonization and named it Avyon, which is the missing 7th sister of the Pleiades, after their home planet in the Lyran constellation. Just as Abraham's descendants were given the land of Canaan as their new home, Devin was given the Pleiades.

Fig. 6

Amelius, the first soul to incarnate
into physical human vehicle, incarnated
on the Lyran Avyon.

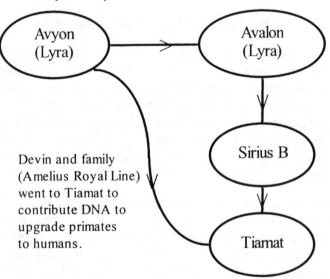

Devin and his extended family, once again, upheld the promise to maintain a pure genetic strain and they married among themselves. They also polarized to the feminine side. After many generations they became stagnant, having repressed the masculine side of themselves. This stagnation was pre-planned by the Founders and Game Engineers when they first developed the Game for the universe.

To jump start their now-stagnant spiritual evolution, the Avyonians would proceed to the next step in the universal polarity integration game plan—the destruction of Avyon by the masculine-polarized Reptiles.

Fig. 7

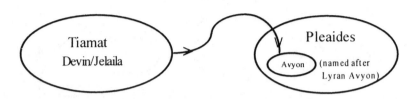

Devin, patriarch of the Amelius line,
took his wife, Jelaila and 88 of his
family and left Tiamat for the Pleaides.

This would create such a large emotional block in the psyches of the Avyonians toward the Reptiles that it would take millions and millions of years to integrate. This guaranteed the extension of the polarity integration game for a long time. This Avyonian block would be transferred to Earth through the Royal House of Avyon on Nibiru. The dissolution of this giant block through integration would shift Earth to 5th dimensional status in the final years of the Game. More on this in a future publication.

Fig. 8

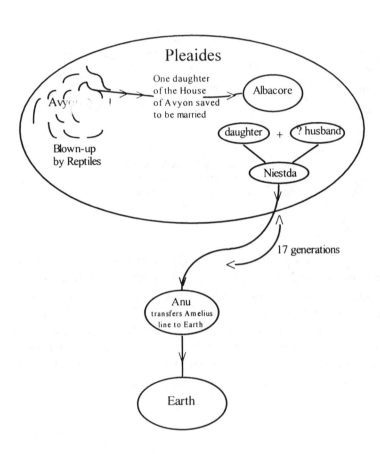

After the destruction of the Pleiadian Avyon, the Amelius line was transferred to Nibiru. This was accomplished by taking a daughter of the House of Avyon to another Pleiadian planet, just before the destruction of Avyon, to be married. The child born of that union was Niestda, the future first ruler/commander of Nibiru.

Seventeen generations later I, Anu, became the ruler/commander of Nibiru. I transferred the Amelius line to Earth through my children. The Amelius Line was also transferred to Earth by Amelius himself, incarnating as Adapa (Adam), though this was not a pure strain as Adapa carried the genes of all four primary universal races.

The two Amelius lines merged in the royal priestly line of Sumer. This occurred through the mating of my family members with the Adapa line through Seth, the third son of Adapa.

Fig. 9

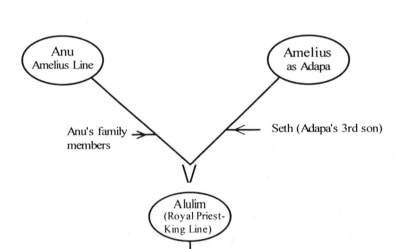

Terah, Abraham's father, was born of this royal priest-king line. The royal House of Judah and David extended from this same line through Abraham's great-grandson, Judah, one of the twelve sons of Jacob. In fact, all mankind alive on the planet today are descendants of one or both of the Amelius lines. Now, back to Tiamat.

The Humans and Reptiles lived in relative harmony on the planet Tiamat for some time, the Reptilian Snakes settling on one side of the Humans and the Reptilian Dragons settling on the other. The Humans were given the arts of farming and animal husbandry. They began growing surplus amounts of food, which they gave to their Reptilian neighbors.

This fostered even more harmonious relations between the two races. It wasn't long before the galactic and universal grapevines began buzzing with the news. The Founders, the planetary Spiritual Hierarchy, and the Etheric Sirians, were beside themselves with joy. It seemed polarity integration was about to occur.

Fig. 10

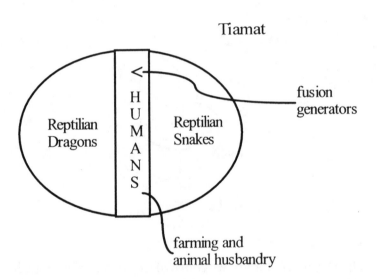

Tiamat

Reptilian Dragons

H
U
M
A
N
S

Reptilian Snakes

fusion generators

farming and animal husbandry

But, their hopes were dashed when a group of Reptilian Orion Council members arrived on Tiamat to investigate. They had heard the news of this harmonious living arrangement and they were not pleased. This was in direct violation of their creation myth. The Humans should have been destroyed. The Orion representatives spoke at length with the ruling families of the Reptiles but could not change their minds.

So, they decided to do their usual thing. They began sowing seeds of mistrust among the Reptiles. They began telling them that the Humans were secretly planning to destroy them and keep Tiamat for themselves. It took the Orion Council members a mere 10,000 years or so to accomplish this.

Fig. 11

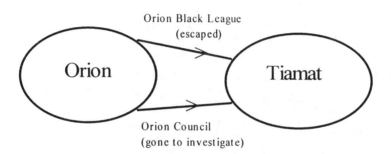

Orion Black League
(escaped)

Orion   Tiamat

Orion Council
(gone to investigate)

The Orion Black League people, who had arrived on Tiamat shortly after the Reptiles, did not help matters. They stirred up the Humans. The Black League did not trust the Reptiles and would have been happy to see them destroyed. Before long, the two races reached a crisis point. The Reptiles agreed to destroy the Humans through a form of germ warfare. The Humans sought the assistance of their parents, the Pleiadian Avyonians, along with the Etheric Sirians and others.

A plan was made. The Humans would leave Tiamat and move in the starship Pegasus to a new homeworld where they would continue their evolution. The Etheric Sirians who had not become Earth Sirians would mutate back into aquatic primates (whales and dolphins) and return to the planet to maintain the biosphere until their brothers and sisters, the Earth Sirians, could once again raise a new race of human vehicles for the Etheric Sirians to use to carry out their land guardian duties.

Nibiru would be sent in to destroy the Reptilian colonies by imploding the fusion generators located in the middle of the Human stronghold. The fusion generators maintained the magnetic and electrical force fields of Tiamat. Once the fusion generators were demolished, Tiamat would be rendered lifeless and 98% of the Reptilian civilizations would be destroyed. The Things (part Etheric Sirian, part animal) would also be destroyed, thus freeing the souls of the Etheric Sirians who inhabited them.

Fig. 12

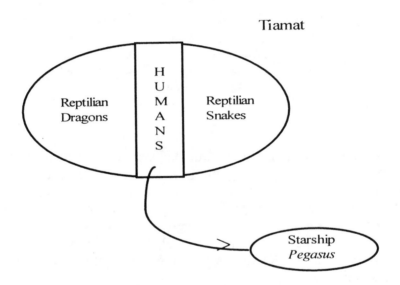

But they, like the Reptiles, would not all be destroyed. A small portion of them would survive by moving underground. It is said old habits die hard, and this was the case with the Etheric Sirians. They would return to the new Earth, and continue in the incarnational cycle of the animals.

The destruction of Tiamat was my first assignment as the new ruler/ commander of Nibiru. I was guided by the Nibiruan Council of the

Galactic Federation in this assignment. I would like to interject again and give you a little background on the Galactic Federation.

The Galactic Federation

The Galactic Federation was formed to address the overwhelming need for an organization that would oversee the many scattered human colonies, aid communication, and foster peace between the different races in the universe. There was too much conflict in the universe, what with the Reptiles and Humans fighting, and they were only part of the chaos.

The Galactic Federation was formed before the creation of Nibiru and after the destruction of Avyon, this being the catalyst for its formation. The Galactic Federation is quite enormous and diverse. Books have been channeled that describe it, but the Galactic Federation is so large and multi-layered that it would take a whole library of books to thoroughly address the subject. One book, in particular, that I had Jelaila read is You *are Becoming a Gaiactic Human,* channeled through our Sirian sister Virginia Essene. I found it to be most helpful in describing the Galactic Federation to Jelaila.

The creation of Nibiru was one of the first major feats of the Galactic Federation. The Galactic Federation realized the need for a battlestar to patrol the galaxy in a peace-keeping capacity. There were many skirmishes with the Reptiles during the time before the destruction of Tiamat, but the destruction of the Reptilian colonies on Tiamat started the Great Galactic War.

This war would last for many millions of years and would involve the entire galaxy. No star system would be left undamaged by the conflict. Please remember, this is polarity integration in action. One must experience both sides before one can integrate. The Great Galactic War provided a perfect game board for playing the Game. All souls in the universe at that time were given the opportunity to incarnate on both sides for the sake of experience, including the Avyonians.

The Destruction of Tiamat

I took over command of Nibiru and headed for your solar system and Tiamat. Tiamat was in, roughly, the same orbit as Earth is now. It was just a little further out from the sun. Being the 12th planet of your solar system, we entered your solar system from the back, as this is our orbital pattern. We passed by Neptune and Uranus and, as we drew closer to Saturn, we let her gravitational pull move us into a position that aligned us with Tiamat.

Fig. 13

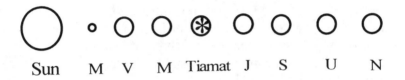

Sun   M   V   M   Tiamat J   S   U   N

We harnessed a satellite of Saturn and propelled it into the middle of Tiamat with such force that it distended her. Then we shot a laser beam into this cleavage at the location of the fusion generators. The beam blew out the force fields, and Tiamat was left lifeless.

During this time I was in constant communication with my superiors in the Nibiruan Council. I was given the order to continue with my regular orbit and, when I arrived back in your solar system, to finish the job. That meant breaking Tiamat in two and shunting the upper half into a new orbit to be rehabilitated. This upper half would become Earth.

Fig. 14

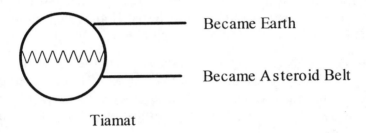

Tiamat

The lower half would be smashed into pieces and become the Asteroid Belt.

The Saturn satellite we had harnessed became Pluto. We shunted it into orbit and established an outpost command there. We use this outpost to monitor events in your solar system.

Fig. 15

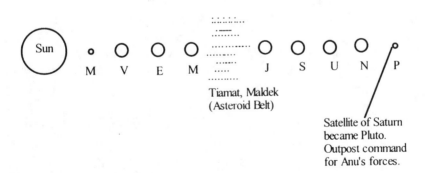

Once Tiamat was put out of commission we didn't hear much from the Reptiles for a while. Tiamat's near destruction had been a big blow for them, and they needed time to recover. Only two percent of the Reptiles survived by, either going underground, or by being taken aboard Nibiru. We offered sanctuary to those members of the ruling Reptile families that had been cast out by the Reptilian Council for not going along with the plan to destroy the Humans.

With the cooperating remnant of the House of Aln aboard Nibiru, I set out for my next assignment-the destruction of the Reptilian Royal Planet Aln in the Orion constellation.

Once the Reptiles recovered from the shock of the destruction of Tiamat and Aln, the Great Galactic War began. They decided that the Humans in this Galaxy had to be destroyed for peace to reign. They armed Maldek, their military outpost planet in your solar system to do just that.

The upper half of Tiamat, now called Earth, was rehabilitated and reseeded by the Felines. They, with the help of the Christos Sirians, seeded for plants, animals, and a new land guardian race. Once again the evolving land guardian race was being watched over by the Earth Sirians. But, this time there was a difference—the Etheric Sirians had their brothers and sisters, the Aquatic Sirians, also on the planet.

Fig. 16

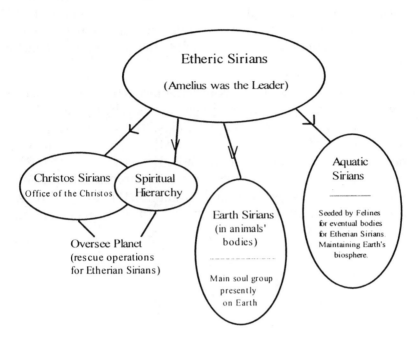

Now we have the Christos Sirians working with the Spiritual Hierarchy to oversee the planet, the Aquatic Sirians maintaining the biosphere, and the Earth Sirians watching over the new evolving land guardian race.

A few million years after the reseeding of Earth, a new Human colony, called Hybornea, was established. The human colonists came from all over the galaxy and Hybornea survived and flourished for nearly one million years. It was a Lyran/Sirian colony and attracted colonists who leaned toward the feminine polarity. The Hyborneans tried to help the Things and managed to free a few of them from the animal cycle, but many still remained to be freed. Hybornea was destroyed in a massive attack by the Reptiles. They launched their attack from Maldek, which we later destroyed. Maldek is now part of the Asteroid Belt, along with the lower half of Tiamat.

The destruction of Hybornea fueled the fires of the Galactic War even more. This would be the final great battle. Nibiru was ordered to destroy Maldek and run the Reptiles out of your solar system for the last time. This I accomplished but not before the Reptiles had destroyed the Human colonies on Venus and Mars, leaving them both uninhabitable. The destruction of Maldek brought the Galactic War to an end.

The destruction of Maldek cost Nibiru her protective forcefield, due to the massive amounts of atomic weapons used by the Reptiles in their attacks. Maldek was destroyed and Nibiru was severely crippled and its people were dying by the thousands. The only way to save the Galactic Federation's great battlestar was to find large quantities of gold to suspend in the forcefield around Nibiru. This would protect the battlestar from the radiation that swept through space. Now, let's move to Earth's history at the time of our arrival, 480,000 years ago.

12

Earth 480,000 B.C. to 100,000 B.C.

At the time of our arrival on Earth 480,000 years ago, the Lemurians, the Things, the Reptiles (the descendants of those who had gone underground at the time of Tiamat's destruction) and the evolving Human primates were residing there. The primates had been seeded by the Felines and the Etheric Sirians and had reached the evolutionary stage of homo erectus when we arrived. They were intelligent, telepathic, and lived in peace with the animals in the wild, in a communal type of society.

There also existed the civilizations of Yu (the Orientals) in Asia and the civilization of Atlantis (the red race). These were started by my cousins who arrived on Earth after the destruction of Hybornea to recolonize the planet. Ashen started the Yu civilization and Alta (Atlas) started Atlantis. Both were considered offshoots of Lemuria, as both had made agreements that the Lemurians would act as a mother empire to them.

Each of the three civilizations had begun in Lyra as white races but had agreed to allow the Felines to change their DNA so they could better adapt to the climate in their chosen areas of the planet. This is how the different red, yellow, and brown races came about. The black race originated from the evolving Human primates. They would become the land guardians once they were crossed with the Nibiruans to obtain the necessary DNA upgrade to homo sapien. At that time they would be able to carry the souls of the Earth Sirians.

The white race was established by my family, the Royal House of Avyon from Nibiru.

We came to Earth for two reasons. First, we wished to acquire gold to put in the forcefield around Nibiru to save our people. Second, we had made an agreement with the Christos Sirians to create, with the help of the Felines, the two-stranded DNA Human body for the Earth Sirians, from a cross between us and the evolving land guardian race.

Alalu, the former Nibiruan commander and my older brother, had previously arrived on Earth and had found gold. I sent my eldest son, Enki, and fifty of my best astronauts (Anunnaki) to Earth to establish mining operations. Time was running out on Nibiru so there wasn't a moment to lose.

Now, allow me to back up for a moment and fill you in on some more history. The Reptilian remnant on Earth had grown into quite a large population, and we knew we would not be able to mine peacefully without some kind of truce and agreement. This was accomplished by my taking the Dragon queen, Dramin, who had resided on Nibiru since the destruction of Tiamat, as a wife. Enki is the child of our union. Enki is half Nibiruan Human and half Dragon, or Reptile. All three of my children and grandchildren had been born on Nibiru and all had reached maturity by the time we came to Earth. These included Enlil, Enki, Ninhursag, Ninurta, Nannar, Marduk, and Thoth (Ningishzidda). (See "Figure 1. Anu's family tree," on page 52).

Earth Mission

Enki's shuttlecraft landed in the sea near Mesopotamia. He and the Anunnaki set up camp and went to work, extracting gold from the water which Alalu had found earlier. Enki built the first city which he named Eridu—Earth Station One. He received much assistance from his Reptilian relatives.

Ninhursag, my daughter and Chief Medical Officer for the Earth mission, arrived soon after to provide medical attention for the astronauts. I arrived a short time later with my other son and heir-apparent, Enlil to assess the gold situation.

The Anunnaki were working as hard as they could, extracting the gold from the water but it just wasn't enough to save Nibiru. Our people were still dying. We began a search for more gold and located it in Africa. But there was only one problem, it was underground and would have to be mined. Tensions had been running high between Enlil and Enki. Enki felt that he should be given Earth since it was he who had arrived first, who had toiled and sweated to build the first camp and city, Eridu, not to mention extracting the gold. Since he had done all the work, he rightfully felt entitled to his due. Enki is the son of Earth's Dragon Queen and, in the Reptile's point of view, heir to Earth as well. The Reptiles claimed Earth as their own, in accordance with their creation myth, and they still do.

Enlil, as the son of my sister/wife Antu, was, by Pleiadian law, my rightful heir. This meant he would be heir of Earth, as it came under my dominion. A quarrel between the two ensued. Both had a valid case. The solution was to have them draw straws, which they did. Enki was given all of Africa as his domain and Enlil received the remainder of Earth as his.

The War of Olden Gods

We moved the astronauts to Africa to begin the mining operation, and Enki went with them to oversee it. I left Enlil in charge in Eridu and prepared to leave for Nibiru, but I was stopped by Alalu's grandson Kumarbi. Kumarbi had been, for some time, harboring resentment against me and my sons and grandsons. He felt he should have been named commander after his grandfather had stepped down. It seemed he had managed to gain the support of the astronauts on the orbiting space station (the Igigi) and now intended to take command by force. Well, this did not happen. My grandson, Ninurta, and many others joined in a battle that ended with the defeat of Kumarbi and his allies. Needless to say, I replaced the astronauts on the space station.

The Garden of Eden

Events on Earth continued to move forward. Enki was sending ample amounts of gold for refining to the newly-built metallurgical center in Mesopotamia, Bad Tibira, and from there to the orbiting space station to be shuttled on to Nibiru. Enlil had set about building four new cities. They were:

Sippar (the spaceport), Nippur (the mission control center), Bad Tibira (the metallurgical center), and Shuruppak (the medical center). My daughter loved that hospital. She had Enlil build it with all the latest medical technology and extensive laboratory facilities.

Life continued to flourish, and the new cities and surrounding area were a lush and beautiful sight to behold. We brought seeds for fruit trees of every kind from Nibiru. Mesopotamia became a virtual Garden of Eden. In fact, it was the Garden of Eden or E. Din, as we called it. Things continued to move along at a brisk pace, and everyone was working and living in harmony. This tranquil growth continued for, roughly, 200,000 years.

The Anunnaki Revolt

Around 250,000 B.C., the astronauts in Africa mutinied. They were dissatisfied with mining gold in the hot interiors of the Earth. They had reached the breaking point and joined together in a revolt. Enki called Enlil to notify him of the situation. When Enlil arrived at the mining operation, the Anunnaki took him hostage. Enki, although siding with the Anunnaki, asked his men to release Enlil, which they did. Enlil accused Enki of inciting and encouraging the Anunnaki and wanted him brought up on charges before the Council. Both brothers returned to Nibiru to speak before the Council.

It was decided that the Anunnaki were more valuable doing the jobs they were trained for than mining gold. Enki believed the solution was to create a worker race to mine the gold, as this would also take care of their agreement with the Christos Sirians. So Enki, being a master geneticist as well as an engineer, along with Ninhursag, retired to the laboratory at Shuruppak to create the workers.

Creation of the Lulus (Primitive Workers)

The creation of the new worker race was the second of two things we were to accomplish in our agreement with the Galactic Federation, the planetary Spiritual Hierarchy including the Christos Office, and the Founders. For this we were given assistance by the Felines. My Feline wife, Rayshondra—mother of my daughter, Ninhursag, was very helpful. She arrived in Shuruppak to oversee the work. As a mother/daughter team, they work well together. Rayshondra was quite experienced in the field of genetics. She had trained Ninhursag, and Enki had been tutored by her as well.

There were others from Sirius A who came to lend their expertise, Natara and Joysia in particular. This was a much-awaited event. Not only for the Nibiruans, who needed to save their planet, but for the Earth Sirians as well. The creation of their new physical vehicles was at hand. The Christos Sirians also arrived to witness this momentous event.

Many on your planet believe we created this worker race just to mine gold and to serve us. I can understand their reasoning, since the more important spiritual reasons were removed from the historical records we had written and left behind for you. I would like to share them with you.

As the descendants of Devin and the House of Avyon, we were karmically and genetically tied to Tiamat/ Earth, although the evolving humans had been taken to a new homeoworld in the Pegasus star system. So, in essence, we were still the parent race for a race of beings; they just weren't on Earth anymore.

With the creation of the worker race we were, once again, the parent race for beings on Earth. This race was the result of the second seeding of land guardians by the Etheric Sirians and the Felines. The date for the completion of the evolution of this second seeding is your present time, and this date could not be met without genetic acceleration of the race's DNA, by us. Without the genetic acceleration, the evolving Humans would not have kept the date, and this would have stalled the spiritual evolution of the galaxy and the universe.

Additionally, we would gain understanding of polarity integration by parenting this new race. This is part of our Nibiruan/Pleiadian Divine Plan. It is our spiritual reason for coming to Earth.

Last, the DNA acceleration of homo erectus to homo sapiens was needed to give the Earth Sirians the time to transfer from the animal bodies to the new Human physical vehicles through a period of incarnations. This transfer took 200,000 years to complete.

The DNA Upgrade

This new upgrade gave the Human primates many new abilities. They had the ability to reason and make more complex choices. They could learn the fundamentals of reading and writing. Though they retained their telepathic abilities, they were not given any new psychic abilities. Those would be given when all the Earth Sirians had made the transition. The other psychic abilities would be needed for spiritual progress. At this time we were concerned with giving the Earth Sirians bodies that were not too complex, as they had severely regressed during their millions of years in the animal incarnation cycle.

As you can see, there was more to creating this worker race than meeting our needs. Although, I admit that those were important too. Nibiru was dying because of radiation and it was still needed as a Galactic Federation battlestar to protect the galaxy. I hope you can now understand why this race needed to be created and why it had to be then.

Let us return, now, to Shuruppak. Nin (my nickname for Ninhursag) and Enki, gathered the necessary tissue and blood samples from the evolving primates homo erectus. This was not difficult, as they were already an integrated part of our society and many lived among us.

Next, they found the appropriate combination of DNA and combined it with DNA from our people. They obtained eggs from the wombs of the primate women and fertilized them with the sperm from some of our astronauts. These fertilized eggs were then inserted into the wombs of some of our female astronauts and lab assistants.

Twelve new children were born, retaining all the abilities of the evolving primates but now, with the reasoning power (spirit) of a galactic human. Before, they had only an animal spirit—albeit a very intelligent animal spirit—now they had a human spirit. But, the soul was still lacking.

You may ask why we did not give the worker race a greater upgrade? Let me explain. As I mentioned earlier, the Earth Sirians had spent a lot of time in the animals. During this time they became accustomed to the simplicity of the animals. In order to transfer them we had to create a body that was, in many ways, functionally similar to that of the animals. Thus the upgrade took them gently to the level of a six-year old human.

It would take some time to convince the Etheric Sirians to transfer into these new bodies. Not all the bodies birthed would contain an Earth Sirian soul. Those without an Earth Sirian soul would be like an intelligent pet that could speak and perform simple tasks.

We continued this birthing method for some time, but after a while our women got tired of being pregnant all the time. You see hybrids cannot procreate on their own, so we had to do it for them. Once again, Enki and Nin returned to the lab to do another genetic adjustment. This upgrade would allow them to procreate on their own. We named this new race the Lulus, which in our language means primitive worker. I realize that some of you give Lulu a different meaning. I am not here to change that, only to state our interpretation of the name. Enki and Nin would return to the lab twice more for upgrades. I'll share that information when we come to it.

The Big Campaign

The transfer of the Earth Sirians from the bodies of animals to humans was quite a challenge. Many did not want to transfer, as they had become comfortable in the animal bodies. They had become used to only having to use animal instinct. Tackling the complex use of a human body would be like your choosing to tackle the use of an intricate computer when you had been used to using a manual typewriter.

The tactics we used could be likened to a propaganda campaign. The Lulus were of great help in this matter, as they could communicate with the Earth Sirians still in the animal bodies. The Lulus helped them understand the benefits of using a human body, as they were proof of the benefits of upgrading. They acted as peers and teachers for the newly transferred souls. Nin was also helpful in getting the Earth Sirians to transfer. She was greatly loved by the Lulus. They called her Mama. As you can see, your term for mother is very ancient.

Nin was the mother of this new race, nurturing and caring for them. Enki was the father. Unfortunately, not all of my people or the Reptiles felt the same way. Many viewed them as free slave labor, which resulted in their mistreatment. This angered Nin greatly, and she, along with Enki, continually besought Enlil, who was Prince of Earth, to give them more rights and protection.

Shortly after the creation of the new race, about 150,000 years ago or so, a new glacial period began, and the Lulus regressed, along with the many other civilizations on Earth. We Pleiadians were not the only ones to colonize or seed Earth. There were others such as the Andromedans. But, all these were here in accordance with the Divine Plan for Earth and approved by the Christos Sirians and the Spiritual Hierarchy. Survival became uppermost in the minds of mankind, so no spiritual or evolutionary progress was made. This is one of the reasons it took 200,000 years to transfer the Earth Sirians.

There were periods of time when it was much easier to be an animal than a human. Humans at that time were considered fair game by the animals, so it made for a precarious life. There was one period when the humans in certain civilizations, such as Lemuria, had to live underground to escape death. This was when the dinosaurs and other large animals were roaming the Earth in packs.

I want to explain here, how the dinosaurs were created. It happened through a form of mutation that occurred when Earth was reseeded.

The additional boost of energy sent to help the new vegetation grow rapidly caused species already grown, to get even bigger. Hence, the dinosaurs and other large animals. These now-extra-large creatures had survived the destruction of Tiamat.

The Anunnaki Marry the Daughters of Men

Around 100,000 B.C. a warming trend began, and the spiritual and evolutionary progress of the humans was on the upswing again. It was during this time that the astronauts were beginning to marry the Lulus. This created some very tall people, because those children carried the gene for height from the Etheric Sirians, who were between eleven and twelve feet tall, as well as our height gene.

It was important to distinguish that the astronauts were marrying the Lulus who had Earth Sirian souls. They were uninterested in marrying Lulus who had only the animal spirit. In essence, the astronauts preferred wives who could reason and intelligently communicate.

We have covered the period from our arrival to 100,000 B.C. Obviously, there is a lot more, but I will leave that to those who have written books covering that period of time. I have asked Jelaila to give you the titles and authors' names so you can find the information if you chose.

Author's note: Anu has asked me to give you titles and authors for information on Earth history, so here they are:

The 12th Planet - Zecharia Sitchin

The Wars of Gods and Men - Zecharia Sitchin

The World Before - Ruth Montgomery

The Prism of Lyra - Lyssa Royal and Keith Priest

You are Becoming a Galactic Human - Essene and Nidle

The Sirius Mystery - Robert Temple

The Pleiadian Agenda - Barbara Hand Clow

Inanna Returns – V.S. Ferguson

Edgar Cayce's Story of the Origin and Destiny of Man – Lytle W. Robinson

Edgar Cayce on Atlantis - Edgar Cayce

Earth - Barbara Marciniak

Flying Serpents and Dragons - R.A. Boulay

Nothing in This Book is True, But It's Exactly How Things Are – Bob Frissell

The Only Planet of Choice - Schlemmer and Palden

Extraterrestrials in Biblical Prophecy – G. Cope Schellhorn

The Sumerian King List - Thorkild Jacobsen

There are many others, including ancient manuscripts such as the Babylonian Epic of Creation, the Mahabharata, the Bible, and many others that contain key pieces of this puzzle. As Devin/Anu once told me, I was to do the research to find the information, and then he would put the pieces together in the proper fashion. This is why I ended up with a whole library of books. Each one had a piece of the puzzle, but I had to read the books to find them.

13

Earth Story 75,000 B.C. to 11,000 B.C.

Another Ice Age occurred around 75,000 B.C., and, once again, mankind regressed to survival of the fittest, although some cultures fared better than others, having attained a greater degree of evolutionary progress. One group in particular, was the Cro-Magnon Man, as you call them. Of all the others, this group survived intact to continue evolving. The other groups that died eventually incarnated into the Cro-Magnon group, giving the newly-arriving souls an evolutionary acceleration, as they had been less advanced than Cro-Magnon Man in their physical evolution when they died.

Around 50,000 B.C. a major event took place. Earth was warming and evolution was in full swing. Enki and Nin were given an order by the planetary Spiritual Hierarchy and Christos Sirians to, once again, upgrade the human bodies. This time it was for the purpose of spiritual evolution rather than physical and mental evolution. Before I begin relating this part of your history, let us take a moment to examine the events around the rest of the planet.

The empires of Yu, Rama, Lemuria, Egypt, and the new Mayan empire were sending representatives to Atlantis to meet in order to determine the best way to destroy the dinosaurs and other large animals that were preying upon the human population. Their solution would see the end of the large animal population and cause the deaths of many people. This was in accordance with Earth's Divine Plan.

The death of these large animals would free the last of the Earth Sirian souls still using animal bodies and would enable us to transfer them for the next step in their Divine Plan—the introduction of the Christos seed for spiritual evolution. This would give them 52,000 years, ten 5200-year cycles, to spiritually evolve into land guardians, capable of managing the planet. During this time they would be fully conscious beings for all but the last few thousand years.

Fig. 17

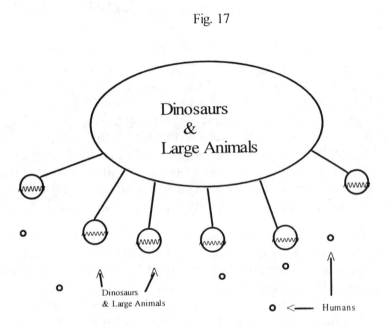

Both on planet, around 50,000 B.C.
at the same time and place!

The Spiritual Hierarchy, the Christos Sirians, the Galactic Federation, and the Sirian A and Nibiruan Councils would convene once every cycle to assess progress and make the necessary adjustments for continued evolution.

The Atlanteans had harnessed the energy of crystals (through the assistance of Marduk, who had given them the technology) and were on their way to fulfilling their (Marduk's) plan for world domination. But this would never come to fruition.

Lemuria was struggling to maintain its Lyran/Sirian form of government and maintain peaceful relations with Atlantis. Many Lemurians had already migrated from Lemuria to other parts of the planet. Many had heard and believed the prophecies of the coming destruction of Lemuria. Some areas to which they migrated were the ones known today as South America, Mexico, America, and Northwest and Central Europe. Now, back to the Christos Upgrade.

After the death of the dinosaurs, or most of them anyway, the transfer of Earth Sirian souls was complete. It had taken a long time, due to the splitting of some of the souls. Many of these Earth Sirians did not want to transfer, so they fragmented themselves, which enabled them to experience life in a human body while still in an animal body. This allowed them the opportunity to sample life in a human body before making the final jump. They were not easily convinced that life in a human body was better.

Adam and Eve (Adapa and Lilith)

The Christos upgrade was given through Amelius himself. As head of the Etheric Sirians, it was only natural for him to be the Adam, or Adapa, as he was affectionately known by us. Adapa was born with Human, Reptilian, Feline, and Carian bloodlines.

This was accomplished by Enki donating sperm to fertilize an egg provided by Nin. This fertilized egg was then inserted into the womb of Ninhursag. Before long, Nin gave birth to Adapa (Amelius).

Fig. 18

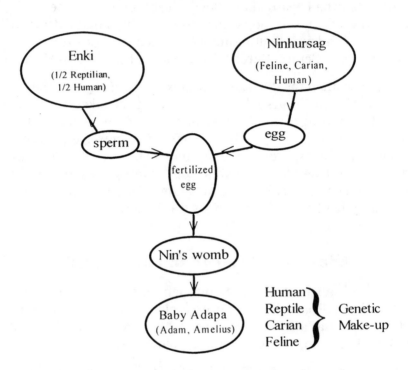

He was a perfect human baby, and through him the integration of the two races would eventually be accomplished. This meant that mankind would carry DNA that was both Reptilian and Human. When we finally accept and love ourselves, we love all aspects of ourselves—Reptilian and Human.

Fig. 19

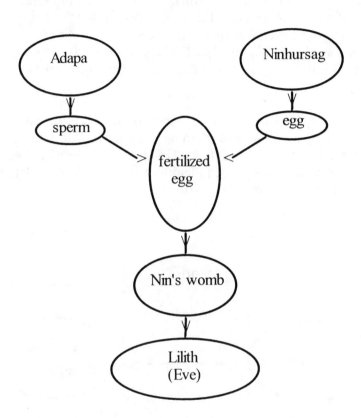

The Feline and Carian DNA was added in the Christos Upgrade to provide an additional boost of Compassion energy from two races that had already accomplished this integration of Light and Dark.

Enki and Nin lavished Adapa with love, just as any loving parents would. Enki taught him everything he knew, and Enki was, and still is, extremely well educated.

When Adapa was two years old, sperm was taken from him and fertilized to birth his other half. We called her Lilith. You called her

Eve. Her birth mother was Ninhursag. Both were fully conscious beings, as were their children. Lillith grew up with Adapa playing in the Garden of Edin—the E. Din Compound. The most important thing about this upgrade was that it would pave the way for the spiritual evolution of the Earth Sirians, now finally inhabiting the bodies of only humans. We also created Adams for the red, yellow, and brown races on the planet and placed them in the chosen locations as directed by the Spiritual Hierarchy. Adapa and Lillith stayed with their parents in the E.Din or as you call it, the Garden of Eden.

Amelius/Adapa, like everyone else incarnating on the Earth plane, agreed to forget who he was when he incarnated. But he started to remember again, just as you have. He and Lilith were married to each other in accordance with Pleaidian Law. Lilith was related to Adapa, as she came from his seed.

Fig. 20

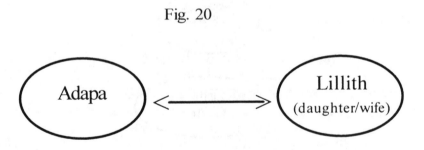

The Biblical Lord in the Garden of Eden was none other than Enlil. The snake, or serpent in the Garden of Eden, was Enki. The snake is Enki's symbol.

Adam and Eve and the Apple Incident

Lilith (Eve) did not give Adam (Adapa) an apple to eat. This was written into the Bible by the church fathers to impress upon mankind the fact that woman, and woman's sin, brought about the downfall of man.

At the time this occurred, the churches were trying to wipe out the matriarchal system, which was world-wide at the time. The Goddess had to be suppressed and repressed. Like the supposed crucifixion of Christ, the apple incident never happened, but was written in by the church fathers to impress upon mankind the idea of sin. The church fathers would then be the only ones who could absolve mankind of sin, through the Blood of Christ.

Let me state here for the record—there is no sin! Jesus Christ was, and still is, conveniently being used by the churches to control mankind. Jesus was just a well-traveled, well-trained Jewish rabbi who returned to his native land to disseminate the knowledge he had gained. His most important teachings were about compassion. He gave to mankind then, what we are giving to you now—The 13th-dimensional Formula of Compassion.

The churches taught that if people did not receive absolution for their sins, they would go to Hell. The Jewish oral tradition also carried the story of the Apple incident, but it was a fruit, not necessarily an apple. This was inserted into their oral tradition around 2000 B.C. by Marduk.

The Tree of Good and Evil (ancient wisdom), was a symbol for the new school Enki had formed to teach the ancient knowledge to Adapa, Lilith and their descendants. It was known as the Brotherhood of the Snake. The fruit was symbolic of the knowledge being taught at the school.

Fig. 21

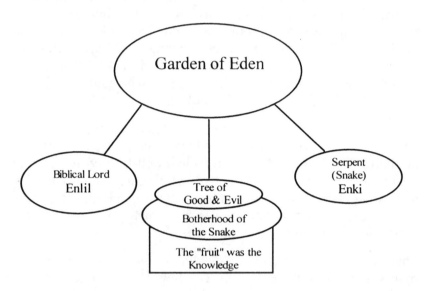

Enki taught both Adapa and Lillith
the ancient knowledge.

Back to the bible story, Enlil was not angry with Adapa and Lilith as the bible states. Instead, he was angry with Enki because he had educated the children of Adapa and Lilith. He was concerned that their future generations would destroy themselves. Enlil felt that that much knowledge in the hands of people who lacked spiritual maturity could lead to the destruction of this new race by their own hand. And there were plenty around to encourage it, in the form of the Reptiles and Marduk. Marduk would use them to further his and his grandmother's plans for domination of Earth by the Reptiles.

Look at your world today, and you will understand how he felt. It could be likened to placing a stick of dynamite in the hand of a child and giving him a match. It has been written that Enlil disliked mankind. This is not so. And Enlil did not evict Adam and Eve from the Garden of Eden either.

He was entrusted with carrying out the Divine Plan of the Earth Sirians, this new race. His desire was to save them from destruction at the hands of the Reptiles, so that they could survive to complete their evolution and take over as heirs and land guardians of Earth. This would then free him and the Nibiruan Pleaidians from the Nibiruans' role as parent race.

The Brotherhood of the Snake

Enki formed the Brotherhood of the Snake to begin the spiritual training of Adapa and his descendants. This school was the forerunner of the Mystery Schools and the later Masonic Lodges. The use of the Apron as a symbol began with Enki and has carried through to your present day. As mentioned, Enlil did not completely approve of this spiritual education of the descendants of Adapa, and he and Enki began to quarrel.

Enki was teaching them universal laws, sacred geometry, and the many techniques for manipulating energy, along with their spiritual teachings. Enlil was aware that the Divine Plan called for slow, spiritual evolution that would stay one step ahead of technological evolution. This would keep the humans from killing themselves with their own technology.

Eventually, the two brothers reached a compromise with the intervention and assistance of the Galactic Federation, the Spiritual Hierarchy, and others. It was decided that Adapa and his children would be moved from E. Din to live on their own outside the compound. Until this time, they had been cared for by us. Moving them would make survival their greatest priority, thus slowing down their technological advancement in order to allow their spiritual advancement to remain one step ahead.

The Brotherhood of the Snake would continue, but only a few would be trained in each generation. This was the beginning of the Priesthood on your planet. Unfortunately, the Brotherhood became corrupted as the priests became power hungry.

In the end, the plan did not work. Enlil was right. By 11,000 B.C., this new race had degenerated to the point where it had to start over. The Atlanteans were the main reason for this decision, due to the influence of Marduk and his Reptilian allies. Their desire to rule the world created war across the planet. The Biblical Deluge was not caused by the astronauts marrying the daughters of man (Lulus), as you have been told. Instead, it was due to Marduk's actions in Atlantis. Let me explain.

The time was around 25,000 B.C. give or take a few thousand years. The Atlanteans, under Marduk's direction, had harnessed the power of a giant crystal. This, at first, was used to power aircraft, ships, and submarines. The Atlanteans were overjoyed at this latest technological advancement. Marduk was worming his way into their good graces and gaining their trust by giving them Nibiruan and Reptilian technology.

He began in the scientific community which then became the dominant portion of their society. Next, he infiltrated the ruling class and soon became the power behind the throne. This led to much dissension within the Atlantean society, pitting the Priesthood and the spiritually-oriented people against the power and technologically-oriented people. There was division in all classes, from the ruling class to the common people. Many Atlanteans supported Marduk and many did not.

The Lemurians sent delegates to the rulers of Atlantis, attempting to warn them of the negative outcome of their quest for world domination under Marduk's direction. Marduk felt his father, Enki, should be the heir-apparent of Earth and Nibiru, not his uncle, Enlil. But, Enki had already given up his quest for rulership. He preferred working in his laboratory, building things, and pursuing his spiritual studies over the daily grind of rulership.

Fig. 22

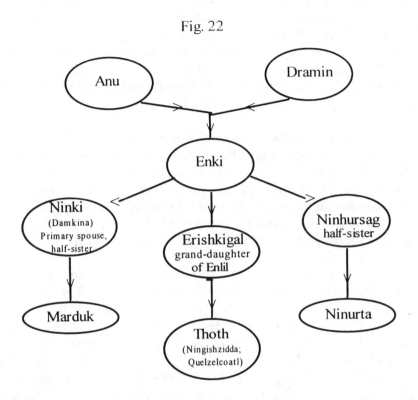

Marduk, on the other hand, preferred ruling and felt he was doubly entitled to do so, since his mother was a Princess of the Snake people and his grandmother was the Queen of the Dragons. Her first husband had died before she married me. Marduk felt that if he couldn't get his father to fight for rulership of Earth, he would. Atlantis was the perfect place for him to begin, as it was far enough away from Mesopotamia and Egypt and from the ever-watchful eyes of his uncle, Enlil, and his father, Enki. In his quest for world domination, Marduk had a very powerful ace—the Giant Crystal. He had managed to harness a comet, one of Tiamat/Earth's ten comets, via a tractor beam from the Giant Crystal, and he used it to threaten the other civilizations into submission.

At any time, Marduk could bring the comet down on any land mass which, because of the comet's size and it's speed of impact, could destroy that civilization. He did this to Lemuria, and it caused the destruction and sinking of the entire continent. He was ready to do it again to the Rama and Yu empires when Nibiru returned to the vicinity of Earth. Nibiru created a momentary disruption of the tractor beam holding the comet in place, and within minutes, the comet came down on Atlantis, sinking the entire continent. This happened at the same time as the Great Deluge, about 11,000 years ago, adding to the mass destruction on the planet.

Marduk was also responsible for the destruction of the Firmament. The Firmament was the band of moisture, about three miles thick, that surrounded Earth. Man did not see the sun or moon until then. Like Avyon, Earth was kept in a subtropical environment, which accounts for the lush, green, garden-like descriptions in your ancient texts. This can be accomplished only if there is a Firmament to filter the sun's rays and provide continuous moisture.

The crystal temples were located underground in various parts of the plantet. They held the Firmament in place. Marduk had Seth, his son, launch the attack on the crystal temples from the Great Pyramid in Egypt. This caused the forty days and nights of rain as the Firmament slowly collapsed. There was as much water in the Firmament as in the oceans. Seth used a laser weapon in the Great Pyramid to accomplish the task. His use of the Great Pyramid for this purpose would later bring about the Second Pyramid War and the emptying of the Great Pyramid of all its equipment by Ninurta.

We held a meeting with the Spiritual Hierarchy, the Galactic Federation, and the Felines, and it was decided not to alert the humans of the impending orbital shift caused by Nibiru's passing. With Marduk instigating war across the planet, we knew it was only a matter of time before mankind was destroyed anyway. Marduk wanted to gain control of Earth, the Pleaides, and the Galaxy. Once he gained control of Earth he was only two steps away from controlling the Galaxy.

All he needed to do was gain control of the MEs—the knowledge-encoded crystals that ran everything. The MEs gave the possessor total power over whatever those particular MEs controlled. As I said, Marduk was after control of the Galaxy. He was not concerned with mankind. They were fodder for his war machine.

His plan was to destroy them and allow the Reptiles to repopulate the Earth. Next, he would destroy the rest of the human population in this galaxy. The Reptiles would finally control all of what they believed was theirs in accordance with their creation myth. Marduk was out to fulfill this goal.

When the time came, we left the planet and moved to the orbiting space station to wait out the flood. This was a sad time for all concerned, as we watched our children, mankind, and over 400,000 years of work destroyed. We had underestimated the Reptiles, their chosen representative, my grandson Marduk, and their desire to own the planet and the galaxy. This attempt at polarity integration had also failed.

What we didn't know was that Enki had told his son Noah (Enki fathered many children) about the coming flood and had him build a submarine to house Noah and his family. Enki sent one of our astronauts to pilot it throughout the coming catastrophe.

I want to state for the record that I am grateful to Enki for defying the Council's decision to keep the news of the coming catastrophe from his son. Without Noah and his family surviving, it would have been much harder to begin again. Enki truly loves mankind and has made many sacrifices through the ages for you. He has returned to you many times at key turning points in your evolution to show you the way. I must also include Nin in this as well.

14

Earth Story 11,000 B.C. to 3200 B.C.

After the flood, we returned to Earth to begin reconstruction. Enki and Enlil brought seed, farming tools, and the knowledge of animal husbandry to Noah and his family. Noah began farming on the slopes of Mt. Ararat where the submarine had settled as the waters receded. Ninurta and Nannar helped by damming and draining the surrounding area as Enki had taught them to do.

By 10,500 B.C., Mesopotamia was reclaimed, along with many other areas. Mankind had once again multiplied and spread out.

The spaceport was rebuilt, but this time it was in a new location, Mount Moriah. Mount Moriah, as you know, later became known as Jerusalem. The other pre-diluvial cities of Nippur, Eridu, and the rest, were also rebuilt. The pyramids survived the flood but had to be dug out from under many feet of sand and muck.

By 9000 B.C. everything was back in full swing. Enki had turned over rulership of Egypt to his descendants, Osiris and Seth. Marduk had been banned from Egypt after his Atlantean disaster. Enki knew he could no longer trust his son. Adad, a son of Enlil, was sent to South America to find gold, which he did.

Enlil had Nannar and Ninurta (his son by Nin) assist him with the management of the rest of Earth. Soon, he too would have to choose an heir. He did not have to adhere to Pleiadian law as this was Earth, and new rules of succession could be made. This was the time period in which the younger generation of Nibiruans began their struggle for power from the older generation. The younger generation consisted of, among others, Ninurta, Nannar, Utu, Adad, Inanna,

Seth, Osiris, and of course, Marduk, though Marduk had been at this game for quite some time. These were my grand-children and great-grandchildren.

It wasn't long before there was conflict on the horizon again. This time it was between Seth and Osiris. Seth wanted to rule all of Egypt and, therefore, slew his brother, Osiris. Horus (Osiris's son) vowed to avenge his father's death and thus began the First Pyramid War.

Around 300 years later the Second Pyramid War began. This time it was the Enlilites, Enlil's descendants, against the Enkiites, his brother's descendants. The conflict was over Enki's descendants controlling the space facilities, namely the Great Pyramid.

After Marduk's Atlantis fiasco, the Enlilites did not find comfort in Enki's descendants retaining control of the space facilities. Their concern was that Marduk would stage another coup and take over the galaxy. The war was ended by the intervention and mediation of Ninhurshag.

Ninhurshag was awarded the space facilities and became known as the Lady of the Mountain in your ancient history. Ninurta emptied the pyramid of its equipment and Thoth (Enki's son by Ereshkigal, who was grand-daughter of Enlil) was given rulership of Egypt, replacing Marduk's line. This occurred around 8600 B.C. From that point on until 3400 B.C. peace reigned on the planet.

Around 3700 B.C. kingship was lowered from Heaven (Nibiru) to Earth. Mankind had finally proven themselves mature enough to rule themselves. The new Priest-King line, half Nibiruan-Pleiadian and half Earth-Human began. This is also when you began keeping time. The calendar was given to you by Enlil, in Nippur.

The Neolithic period, as you call it, began, and Earth had its first half-Nibiruan ruler, Alulim. You would call him a demigod. Until that time all civilizations had been ruled by beings from other planets.

By 3400 B.C., the peace was broken by Marduk. He was up to his tricks again. This time he convinced the Babylonians to create their own spaceship and launching pad (the Tower of Babel). Marduk, as ruler of Babylon, oversaw the construction. Enlil caught wind of it and stopped him. It was decided by the Spiritual Hierarchy, the Galactic Federation and the rest, to confuse mankind's language. This, we believed, would once again slow down their technological progress and delay Marduk from reaching his goal of world domination. He could not do it alone. He had to have assistance from mankind to do the grunt work.

With their language confused they could not communicate, and this was a source of great irritation to Marduk. He had to spend many years teaching them each other's language in order to have them work together on the same project.

It was also during this time that the decision to change the humans' DNA structure to suppress their psychic abilities was put into action. Enki and Nin went back to the lab in Sharrupak to accomplish this directive. It was felt by the Spiritual Hierarchy and others that disconnecting ten of their twelve DNA strands would further slow them down. They had already witnessed the speed at which mankind could progress with the aid of Reptilian technology.

Enki and Nin unraveled the DNA strands and placed implants in the astral bodies to keep the strands from re-fusing. Next, they disconnected those ten strands from the endocrine system in the physical body which stopped the creation of a chemical that activates the pineal, pituitary, and hypothalamus glands. Those glands then atrophied from non-use.

Only a few humans would retain use of these glands in future generations. They would carry a special gene for this purpose. It was decided that a small portion would have to be able to communicate with us, in order for mankind to stay on course with their spiritual evolution. These individuals were prophets, mystics, shamans, and psychics. Mankind was also left with the ability to activate the glands but it would take a truly dedicated person to do so.

With his plans thwarted, Marduk returned to Egypt and set about deposing Thoth. This he accomplished. Thoth went to South America and began the civilizations there. The year was 3113 B.C. He was known to them as Quetzalcoatl, the White Plumed Serpent. As Enki's son, he also carried the serpent as his symbol. Enlil and his descendants carried the cross. The cross symbolized Nibiru and the house of Anu.

Once again Marduk had to pay for his actions. This time he was imprisoned in the Great Pyramid for indirectly causing the death of Inanna's husband, Dumuzi, during his fight for rulership of Egypt. Inanna was screaming for justice and if it weren't for Nin's intervention he would have died there. After his release he once again went into exile.

Around 2900 B.C., Inanna was given rulership of a new colony in the Indus Valley. The location was your modern-day India. All historical records there, regarding the Nibiruans, were given by her to be written down. This was the time when she came into her own. Six-hundred years later she fell in love with Sargon and they built a new empire together. It would be called the Akkadian Empire. Also, during this time, she and Marduk fought bitterly back and forth. The stories of their many battles are recorded in your history books. I will not go into them here.

During this time frame Marduk was working on his newest and latest plan for world domination. This time it involved the building of a massive army on the deserted planet, Mars. This he accomplished and, with his new army, Marduk deposed me as ruler of Nibiru by taking the MEs of Earth, and thus, he became ruler/ commander. But, his victory on Earth did not come without a price.

It was decided by the Nibiruan Council to use a plutonium bomb to wipe out the space facilities and other critical areas. The time line of this action was 2024 B.C. This was accomplished. It was the only way to stop Marduk from gaining control of these facilities and going after control of the entire Pleiades star system. The Council chose Abraham to carry the bomb stored in Sumer to the spaceport for detonation by Utu, son of Enlil and a commander in charge of the spaceport. The bomb caused the destruction of all the space facilities along with the cities, Sodom and Gomorrah, which happened to be in the vicinity of the space facilities. This left a once-lush subtropical area of the planet a desert, and Sodom and Gomorrah at the bottom of the newly formed Dead Sea.

The radiation from the bomb swept in a wind to the cities of Sumer, killing the inhabitants and desolating the surrounding area. It would be many, many years before the area would be fit for human habitation.

Abraham, born of a royal priestly Nibiruan family, was given this area as his legacy. It was later known as Canaan. Much later his descendants would reclaim it from other tribes, with the help of another Nibiruan descendent, Moses.

Marduk was now in control of Nibiru and your planet. He set about changing many things. It was at this time that women fell in stature and were considered the lesser of the species, along with children. Marduk would begin the churches, to stamp out the Goddess and the Pleiadian way of communal life which we were given by the Founders of the Universe. Women who were leaders in the communities were stamped out by being branded as witches and burned at the stake. This process continued through the Dark Ages, and ended in the latter part of your 18th century, A. D. It was the most important thing Marduk would have to do to achieve absolute control of mankind.

Second, Marduk would set himself up as the God among gods. Later, this would be changed to God. He would rule through fear, and this he has done down to your present day.

Please, I urge you to remember: Marduk, the 4th dimensional aspect of Satain, is part of both Royal Houses of Avyon and Aln, and therefore, family. Marduk agreed to play his role as leader of the Dark Forces in order to give mankind the necessary opportunities for spiritual growth. Without his effort on mankind's behalf, mankind would not be able to evolve, as there would be no negativity. Marduk is ready to come home. As you know, it is much harder to play the role of the villain than it is to play the role of the hero.

Here is where I end this story of your history. There is much information of that time from 2000 B.C. to your present day. That will be shared in another book. Marduk will be stopped, and it is you who will stop him—not with force but with love, through integrating the dark side of yourselves and, thus, Marduk and his forces.

Take back your power through recoding your DNA. Now is the time. You, the Etheric Sirians, are heirs to this planet. Claim your birthright given you so long ago by the Founders. We, the Nibiruan Avyonians (Pleiadians), your parent race, your brothers and sisters, the Aquatic Sirians, and the Christos Sirians, are here to assist you. You have only to reach out and ask.

Blessings to you, my beloved children of Earth.

Anu, Former Ruler/

Commander of Nibiru of the Royal House of Avyon

Part Three

DNA Recoding

By Joysia and Jelaila

15

DNA Recoding, Reconnection, & Activation

By Joysia and Jelaila

Greetings, beloved ones of Earth. I am Joysia, Chief Genetics Engineer from Sirius A in service to the Nibiruan Council of the Galactic Federation, to you I bring wonderful news! You are now ready to:

1. Activate the production of the Hormone of Compassion.

2. Receive the realignment and fusion of your twelve strands of DNA by removing the implants from your astral bodies.

3. Reconnect those twelve strands into your physical bodies' endocrine systems through your twelve crown chakra crystals.

4. Permanently access your Oversoul or download your Oversoul into your physical body. (Your Oversoul arrangement depends on your contract.)

5. Activate your MerKaBa, crown chakra crystals, and dormant psychic glands, thus making you a fully conscious being existing in a 3rd-dimensional body!

"Fully conscious" means being able to receive transmissions 24 hours a day, from incarnate beings in the dimensions beyond 3rd dimension. You will know what we mean when we say *living in 3rd dimension but not being of it.*

We are very excited to be of service to you in this long-awaited endeavor. Through the process of DNA Recoding, Reconnection and Activation (RRA), you will regain the full powers of a multi-dimensional human. Ten of your twelve DNA strands were disconnected long ago by your Pleiadian parents, the Nibiruans, so that your spiritual growth would catch up and move ahead of your technological growth, thus removing the possibility of your destruction by your own hand.

To the starseeds and walk-ins, yours is a path of awakening. You came from the future back into the past to affect the outcome of Earth events and assist the heirs to the planet—the Earth Sirians. This, of course, means accepting your soul contract for this lifetime and fulfilling it.

For the rest of humanity, which also includes the Lightworkers who are not starseeds or walk-ins, this is your first opportunity to obtain full power. As stated in the writings of Anu, earlier in this book, the rest of humanity is comprised of those beings who belong to the Etheric Sirian soul group who became stuck in the animal incarnational cycle when Earth was still Tiamat.

They needed a way to slowly evolve out of this cycle to the state of spiritual maturity that would enable them to take over the duties of guardians of your planet. We call them the Earth Sirians, the rightful heirs of Earth. The Earth Sirians, too, must accept their soul contract and be willing to fulfill it to be recoded, reconnected, and activated.

In August, 1987, of your years, a number of you came together for a ceremony you named Harmonic Convergence. It was watched and celebrated by all of us here at the Galactic Federation. We rejoiced as you came together around the world to open yourselves and Lady Gaia (Earth) to the higher dimensions.

We knew then that it would not be long before you would be ready for the next major step, DNA Recoding, Reconnection, and Activation of your MerKaBa. Yet, we knew we could not begin at that time due to the level of density still existing on Earth and in your physical and emotional bodies.

We convened a series of meetings, and it was decided to wait to bring this process to your awareness until you had cleared enough density to enable a major portion of you to begin and complete the RRA process. We were well aware that if you began too early, the plan would fail and you would experience so much pain from the density that many of you would leave your bodies and never return to complete your work on Earth.

We sent energy shifts in the months of January, February, and early March of 1996 to shake you up emotionally so you could release some of the emotional negativity stored in your bodies. I am proud to announce that on March 21, 1996 at the Spring Equinox, enough negativity had been released to increase your planetary frequency to the degree needed to allow you access to the higher dimensional RRA process.

We began working through our Emissary Jelaila first, to anchor the frequency of mankind's recoding into the high heart chakra of Earth at Missouri and Kansas. This created the template and made it easier for others to move through the nine levels of RRA.

We are happy to say that as of this writing, Jelaila has completed all nine levels of recoding and reconnection. Jelaila is now fully conscious and multi-dimensional, meaning able to hear and see in the dimensions beyond your 3rd dimension. Jelaila is the 9th dimensional Oversoul downloaded into Joscelyn's 3rd-dimensional body. Her MerKaBa is activated along with her Crown Chakra Crystals. The physical proof can be seen in her aura photos.

Jelaila created the template by moving through the nine levels of recoding, one after the other, then reconnection, and activation. After all, that was her agreement with us—to create the RRA template to place in Earth's grid system.

The experiential information she provided allowed us to create techniques to make the process easier for those following in her footsteps. Therefore, Jelaila has experienced the full range of feeling and emotion in each of the nine levels, from extreme joy to extreme depression, confusion, and pain. But do not feel sorry for her. She will be greatly rewarded for this work she has accomplished when she returns to us. Jelaila will also receive many rewards while on your Earth plane. One of them will be her life mate.

Some of you will move through the nine levels more easily than others. It all depends on your frame of mind, the amount of clearing you have done, and where your lessons lie. In any case, when complete, you will be reconnected to all dimensions and you will begin receiving communications from beings in these higher levels. It will be a thrilling and heart-warming experience.

Many of these Higher Incarnate Beings have not ever been able to communicate with your dimension. There were no channels with a frequency high enough to reach and connect with them. Now, as more and more of you raise your frequency through RRA, you will be able to establish contact with these Higher Incarnate Beings. They are waiting in line to communicate their messages of love and support to you, dear ones.

Many of you may be wondering about the other processes of recoding and MerKaBa activation you have tried. You may be wondering if you are already recoded, reconnected, and activated. My answer to you, dear ones, is this. If you can see and hear clearly into the dimensions beyond 3rd, and you can see the antennae of your MerKaBa in your aura picture, you are already multi-dimensional.

We have sent many to give you processes that would bring you and your planet to the necessary frequency that would allow you to access this higher dimensional knowledge—RRA. You will find the tools we give you now to be the tools that complete the job of recoding, reconnection, and activation. Every process given before March 21, 1996, was given to prepare you for the tools you are receiving now. In other words, they paved the way, and did the job quite well, I might add.

This is the big push, the major production number, the worldwide campaign as you would say. The Nibiruan Council and I are giving you the two *most powerful tools available on your planet for shifting yourselves and Earth to its eventual 5th-dimensional state!*

When enough of you have completed the process, the increased mass frequency you create will shift your planet into higher planes of the 4th Dimension, thus averting a catastrophic pole shift by the battlestar/planet Nibiru!

In closing, I wish to say again how honored I feel to be of service to you at this time in your history. It is, indeed, a most exciting time for you. Each of you will be assigned a genetics engineer of your own, yet please call on me at any time if I can be of assistance.

Love and Light to all of you and a pleasant recoding, reconnection, and activation experience.

Joysia, Chief Genetics Engineer from the Galactic Federation's Sirian A Council

DNA Recoding, Reconnection, and Activation by Jelaila

When I began the RRA process I had no idea of what I was asking for. I remember sitting in my favorite chair, the one with the flowers, that reminds me of an English Victorian cottage. I must have had one in my Victorian lifetime. Anyway, as I sat in this chair in the wee hours of the morning, talking with Devin, I said out loud "Devin, I want to be able to hear you and see you like I did before I came here."

Little did I know I had just asked for DNA recoding and that they had been waiting for this moment for over three years. This was in late January, 1996. I don't remember the exact date.

But, I look back now, having completed RRA and I see just how much my life has changed. I feel I am so much more than I used to be.

I went through the process sequentially to be able to understand the difference between the different levels and then document them. This provided written support for those recoding after me. Now, Joysia and the Council have given you the Formula to use in conjunction with the RRA process.

Joysia says each individual will still move through all nine levels, but now it will go faster. Instead of taking eight months, it will take the average individual six weeks to two months to complete the RRA process. The down-side is emotional intensity. A clear, complete explanation of the 13th-dimensional Formula of Compassion will be given first. Refer to it often as you read through this material.

I encourage anyone desiring to go through the RRA process to first do as much emotional clearing and internal cleansing as you can before beginning. The process will be much more comfortable. The most important step you can take to clear anger is a liver cleanse. The Council chose the liver cleanse they wanted me to use and I must say it worked superbly.

I spent nearly four years in clearing and releasing, with the last year being the most intensive. I didn't know at the time that it was for the RRA process but now I do, and I'm glad I did it. Once again, the Nibiruan Council was looking out for me. They knew what they were doing.

There is much more to say, but I wish to share it with you as we go through the RRA process. I will give you a brief description of each level and then share my personal experiences in each, along with tips on how to use the Formula to expedite your journey through the levels.

Please understand that your experience will be different than mine, and, hopefully, not as intense. I did not have the 13th-dimensional Formula of Compassion (the Formula), as you have it now so my process was longer. I used the steps of the Formula but not in the order that accelerates the RRA process.

Before I begin the level-by-level explanation, I will give you an overview of how the 13th-dimensional Formula of Compassion works in conjunction with the RRA process.

16

13th-dimensional Formula
of Compassion

How the 13th-dimensional Formula of Compassion Works

The first five steps of the Formula bring a particular lesson to your conscious mind. The lesson information can be accessed by you through your higher self or through your guides. The information comes down through the sixth and seventh chakras (crown and third-eye chakras) and arrives in your conscious mind as thought forms, with pictures and emotions.

The emotional blocks that pertain to the lesson are stored in the emotional body and look like dark areas of energy. Some are very large and some are small. The size depends on how long you have been adding emotional energy to the area. The energy is stored in layers like the layers of an onion. The first five steps allow layers of the block to peel off and filter into your physical body through your five remaining chakras. The layers move from the emotional body into the corresponding area of the physical body.

The final four steps of the Formula move the layers of the block being cleared through the body and into the new high heart chakra. Once in the high heart chakra, it is transmuted and released. The hormone used for transmutation of negative energy is the chemical equivalent of Compassion. The new high heart chakra was activated in all souls in this galaxy by a Galactic Proxy, on November 3, 1996. Before we begin, let me give you a few instructions for using the Formula.

How to Use the 13th-dimensional Formula of Compassion

1. Use the Formula in the order it is written.

2. Use the Formula on all issues, both past and present.

3. Use the Formula to release issues as soon as they come up. This keeps you from storing negative energy in your body, thereby maintaining your frequency at its present level. Storing negative energy (emotion), lowers your frequency.

4. When working on an issue involving another person, do the Formula on yourself first, then do it on the other person. Release and find compassion for yourself first, then you can release and find compassion for the other person. This also applies to situations like a car problem, major house repair, etc.

5. Use the Formula to begin Levels One and Two of DNA Recoding. Make a list of all individuals/situations that still cause you anger or any other negative emotions.

DNA Recoding is a process of re-fusing your 12 DNA strands through the removal of the implants in your astral body. *Implants are removed as you clear layers off the emotional blocks in your emotional body through the use of the Formula. As* I said before, blocks look like onions and are removed in layers. You will find that you need to use the Formula over and over on the same issue because each issue consists of many layers.

The majority of the layers may be in past lifetimes and some may even have entities attached to them. Each block is a lesson that you have tried to learn lifetime after lifetime and each lifetime/incident it is not learned, another layer is added to the block. So, don't be surprised if you find yourself learning and releasing the same lesson again and again.

You may find that you have contracts with *many people* for the same lesson. Example: You have contracts with your boss, wife, brother, and best friend to learn the lesson of holding your own power.

You may also find you have *many contracts* with the *same person* lifetime after lifetime because you are both working on learning the same lesson. Example: You have had numerous contracts with your mother, of this lifetime, to learn the lesson of unconditional acceptance. Let us use the example of a conflict with another person to explain the Formula.

13th-dimensional Formula of Compassion

Step One: Lesson
Question: What is the lesson I wanted to learn regarding this person?

Ask your guides to help you on this one. Ask them to show you the lesson you wanted to learn. It will be on your life blueprint. Your life blueprint is your road map through your present lifetime. It contains all the lessons, contracts and major events for your present lifetime, along with the people involved.

Step Two: Contract
Question: What is the contract I made with this person?

Ask your guides to show you the contract(s) you made to learn the lesson. If using the Formula to release one individual, ask for the contract that pertains to you and that individual. There are usually many contracts with many people to learn the same lesson. The ratio of contracts to lessons varies depending on how long and how many lifetimes you have been trying to learn that particular lesson. The more lifetimes, the more present-lifetime contracts for that lesson.

Remember, no one agrees to make a contract with you unless they, too, need to learn the lesson. The other person in your contract is there to learn the flip side of the lesson.

Step Three: Role
Question: What is the role this person is playing to act out his/her part of the contract?

Ask your guides to help you see and understand the role you play and the role the other person in the contract is playing. Ask your guides to assist you in understanding how the roles look as they are being played out.

Step Four: Aspect

Question: What is the aspect of myself this person is reflecting back to me?

Ask your guides to help you see and understand the aspect of yourself the other person or situation is reflecting back to you. They reflect an aspect of you by acting out that aspect in his/her role.

Step Five: Gift

Question: What is the gift this person is giving me by playing their role?

Ask your guides to help you see and understand the gift the other person is giving you by playing their role. The gift is the lesson learned.

Once you understand the first five steps, you should be feeling a surge of compassion and gratitude for the person or situation involved in the contract. If not, ask your guides for assistance in helping you gain this understanding.

The last four steps are used to release the emotional negativity from the physical body, through the high heart chakra, where it has been transmuted by the hormone of compassion.

Step Six: Acceptance

Question: Can I accept the role the other person has played, along with his/her actions, to help me learn this lesson?

Acceptance is one of the four elements of unconditional love. Acceptance is part of compassion and is unconditional love in action. This also includes acceptance of who the person is, without judgment.

Step Seven: Allowance

Question: Can I allow myself to let go of any anger towards the other person who played the role that helped me learn the lesson?

Allowance is also one of the four elements of unconditional love. Allowance is part of compassion and is unconditional love in action. This includes allowing the person to be who they are and follow their chosen path, regardless of how you feel about it.

Step Eight: Release

Question: Can I release this person from blame?

This one is easy when you understand that you are not a victim, but an active participant in a lesson and contract you helped create. Taking responsibility for your part in the contract enables you to release the other person from blame for the role he/she played in the contract. You quickly come to realize that you are not a victim and they are not the villain. Devin has always said that it is harder to play the role of the villain than the role of a hero.

Release is different from forgiveness. Forgiveness implies sin and there is no sin. The other person was only playing a role and therefore, does not deserve to be judged for assisting you in learning a lesson. Sin is a tool created by Marduk, leader of the Dark Forces, and disseminated through the churches, to keep mankind in a state of unreleased emotional negativity. This negative emotional energy is used by Marduk to maintain the fear-based systems, which were created to give mankind the necessary negativity to integrate.

Release is a key element in the Formula. You should begin feeling the release through the high heart chakra by the time you reach this step. The release is created by your compassion, gratitude and appreciation for the other person. If not, you have not grasped the full understanding of the higher dimensional aspects of the lesson. Go back and ask for your guides' assistance to gain this understanding and compassion.

Step Nine: Kindness
Question: Now that I have released this person, can I be kind to him/her and if so, how can I do it and when will I do it?

This is really a three-part question and must be answered in its entirety. The completion of Step Nine finalizes the release of the other person. At this point you should be feeling the intensity of the release through the high heart. I find the degree of feeling differs according to the emotional intensity of the issue. The more emotionally charged the issue, the more intense the release.

I have found this, as have others, to be the most emotional step. I find my gratitude for the other person is equal to my compassion for them. Devin says that those who love you the most are the ones who play the darkest roles in your life. He says that when we chose the lessons we wanted to learn in this lifetime, our loved one stepped forward and volunteered to play the dark role, thereby doing what he/she could to make the role as easy as possible. They do not want us to suffer any more than we have to. It takes a lot out of a person to play a dark role for us. Their love for us matches the darkness of the role they play for us.

Once you feel the gratitude and compassion in your high heart chakra and the release of the blame and anger, you are near the end of the process. The final two steps are to choose how you will be kind and when you will do it. In my story you will see how I did it. I encourage you to complete the last two parts of Step Nine as quickly as possible. The process of clearing the layer(s) will not be complete until you have done them.

17

Overview of the RRA Process

Using the Formula

The Formula enables you to accelerate the RRA process. Here's how it's done.

By using the Formula on the list of people and situations you have issues with, you begin peeling the layers off the emotional blocks located in your emotional body. Your frequency rises as each layer comes off. This moves you through *Level One* and *Level Two* and so on.

The layers of your emotional blocks peel off and funnel down through your chakra system and into your physical body. At this point they will naturally move into the high heart chakra where they are transmuted by the new Hormone of Compassion and released through the high-heart. The physical and emotional release you feel when using the Formula comes from the transmutation of the negative energy layer by the Compassion Hormone and the eventual release of the transformed negative energy from the physical body through the high-heart chakra.

As you raise your frequency, you bring forward those lessons you need to learn in each level. This happens very quickly now, due to the new templates created in Estes Park, Colorado, by the five individuals who completed the RRA process there. The frequency in Estes Park is higher than the frequency in Kansas City. The completion of the RRA process, using the new Formula we had learned about only a week earlier, created a template for an accelerated version of RRA.

Your implants are in your astral/light body, and your blocks are in your emotional body. Both are connected to your physical body. When you use the formula, all three bodies are cleared. First the emotional body, then the physical body and then the astral/light body. Your astral/light body is your MerKaBa. As implants are removed from your astral/light body by your genetics engineer, the astral body becomes lighter and its frequency rises.

Once you reach *Level Three, you* begin going aboard ship to remove the deactivated parts of your implants. You deactivated them when you removed a layer off an emotional block in your emotional body. As the deactivated parts of the implants are removed, your genetics engineer is able to begin re-fusing your DNA strands.

You do not need to be completely cleared of all the layers of your blocks in your emotional body to be able to re-fuse your DNA strands. In fact, you do not have to be completely cleared to be multi-dimensional. That is perfection and we are not going for perfection. The goal is to clear enough to become multi-dimensional again and to shift the planet into the 5th dimension. Perfection would reunite you with Divine Creator, which is a wonderful goal, but one that can be accomplished in future lifetimes.

In *Level Four* you begin seeing clairvoyantly again because you are still using the Formula to release issues as they come forward. Lessons in the form of issues and situations will continue to come to you throughout the recoding experience.

Level Five is where you access the information of your contract for service in this lifetime. Once again it is reached by increasing your frequency through clearing the layers off the blocks in your emotional body. And, of course, this happens because of the lessons that are brought forward in the manner I mentioned earlier.

Level Six sees you regaining your power because you have released layers off blocks that pertained to your powerlessness. All blocks contain issues of powerlessness since fear, which is the basis of emotional blocks, is also the basis of powerlessness.

When you reach *Level Seven,* you have the opportunity to do a major clearing of layers on your blocks. This major clearing gives you the additional frequency boost needed to move you to just below the activation frequency of your MerKaBa. An example would be an increase of 13 degrees instead of the regular 2 or 3 degrees of the other levels.

By *Level Eight,* 10 of your 12 strands are usually connected in your astral/light body and you begin to experience a sense of lightness. People are drawn to you just to spend time in your energy. Your frequency feels wonderful to them. Just like people were drawn to Jesus, they will be drawn to be around you. Setting healthy boundaries is a necessity now in order to preserve your energy and still have some to give to others.

By the end of *Level Nine,* all 12 strands are realigned and refused, and you are ready to have your DNA strands reconnected to your physical body's endocrine system, through your crown chakra crystals, by your genetics engineer.

The chart ("DNA Recoding and Reconnection Chart" on page 135) given to us by Joysia, will be used by your galactic DNA coach to make certain you have cleared enough on each level to complete your reconnection. The reconnection will be comfortable as long as you have cleared enough layers from your blocks. The energy coming through your DNA strands, should you attempt to connect before you have cleared enough, could cause you physical pain in the corresponding area of your body.

If a box on the chart cannot be filled in, it is because there is a core issue that has not been cleared. This is usually one from early childhood or a past lifetime that is so deep you cannot reach it from a conscious state. We have been given a technique by the Council for regressing a person into the block to release it. Many will be trained as coaches and DNA recoding counselors to take a person through the entire RRA process.

Once you are reconnected and all the boxes on the chart are filled in, your Oversoul is either down-loaded into your physical body or a permanent communication line to the oversoul is created. You are ready for activation. *Activation of the MerKaBa allows your Oversoul to begin functioning from a conscious state and allows it to overlay your tapes.* Since you have cleared most of the negative energy from your tapes, your Oversoul will have an easy time overlaying.

Many find there is an intense 30-day period during which lessons continue to come forward. This time helps complete the final clearing of enough layers to bring your frequency up to the degree that is needed to activate your MerKaBa.

It is also during this time that you will need to rescind your Reptilian Vow to break the frequency of 3rd dimension. This too, will be done by a trained counselor. After you have filled in all the boxes on the chart, cleared the core issues, rescinded the Reptilian Vow, and downloaded your Oversoul or gained permanent access to it, you will activate your MerKaBa.

Your genetics engineer will then perform one final operation—the sealing of the "Mouth of God" located at the back of your neck. This will break the frequency of fear that all mankind is connected to, and you will begin feeling as if you are living in another world even though you are still in 3rd dimension.

Your Oversoul takes over and you begin to see and experience life from a higher perspective—the perspective of your higher dimensional Oversoul. You will no longer harbor the debilitating physical and emotional effects of negativity. This does not mean you will not experience them in the moment. It simply means you will automatically use the Formula when situations arise that bring with them these negative emotions. Therefore, you will maintain your frequency by releasing the person or situation and the emotions that come with them, as soon as you can, after it happens. Holding negative energy after activation can cause physical discomfort, as it is too dense to stay in your lighter body.

I encourage you to have an aura picture taken before you begin the RRA process and after you have finished. You will see not only a dramatic difference in the picture, but you will also see the antennae on your MerKaBa. The clarity and glow of your facial features in any further aura pictures will reveal how far you have moved above the 3rd dimensional veil.

DNA Recoding and Reconnection Chart

Name: _____ Date: _____

	1. Courage	2. Focus	3. Sexual Balance M/F	4. Health	5. Bliss	6. Personal Truth	7. State of Grace	8. Personal Boundaries	9. Communal Harmony	10. Knowingness within the moment	11. Balanced Expression of Power	12. Unconditional Love
Pineal												
Pituitary												
Thymus/Thyroid												
Adrenals												
Heart												
Hypothalamus												
Gonads												

The Nibiruan Council, www.nibiruancouncil.com
email: info@nibiruancouncil.com

Life becomes much more pleasant and the need to sleep lessens. Before I begin the complete explanation, let me give you the 1,2,3s of how to start the RRA process.

1. Begin by making a list of all the people with whom you still hold any negative emotions.
2. Do a liver cleanse to give you a jump start. The liver cleanse removes many layers off your blocks, all at one time. Not doing a liver cleanse only prolongs the RRA process since it will take longer to remove the layers.
3. Continue using the Formula all through the RRA process since it will heighten your frequency and move you from one level to the next.

That's all there is to starting the process. Now, let's move on to the complete explanation of the RRA process.

Level One: Releasing Anger

We all begin recoding with *Level One*. This level involves the releasing of old anger stored in your body. You see, anger will actually cause you physical pain, and it is a major detriment to recoding, which takes you and your body from a fear/anger-based reality to a love/compassion-based reality.

This pain is a very effective tool that helps you understand how stored anger affects you physically and emotionally. The pain pinpoints where the anger is stored. For example, if you have abdominal pain, this is where a block of anger is being stored. It is known that anger is stored in the liver, but it can also be elsewhere in the body. Liver cleanses are recommended in this level to release layers of anger and prepare the body for recoding.

The Council and Joysia are adamant about releasing anger and clearing the body of as much density as possible before the implant removals begin. They tell me that the densest emotion is anger and that's why they are targeting it as the number-one emotion to release, not only from the physical body but, more importantly, from the emotional body.

They say clearing the emotional body is the most critical work, since anything left there is manifested in the physical body. So, in other words, you can release all the anger in your physical body, but it will return if not cleared from the emotional body. This is why releasing those individuals connected to the anger is so important. This can be done either in person, by phone, or by contacting the Higher Self of the individuals involved. Use the 13th-dimensional Formula of Compassion to complete this process.

The benefit of *Level One* will be an increase in your frequency, because you will have used the Formula to release and clear your anger issues with others.

My Experience in Level One

When I began *Level One* in late January, 1996, I didn't know it was *Level One*. In fact, I didn't start calling it *recoding* until the first of March. That was the first entry in my journal where I called it recoding. I remember feeling a great desire to make peace with everyone in my past and in the present. I was sitting in the office during lunch one day, and began making a list of all the people with whom I still held anger.

I started with Joscelyn's parents. I called her mother and told her I loved her and was sorry for the years I had blamed her for my problems. We had been in an on-again off-again war since I walked in. I had gone to therapy and was trying to get her to validate my feelings regarding the emotional and psychological abuse of my (Joscelyn's) childhood.

I finally realized that she couldn't see it as I saw it. She didn't understand relationship patterns. Mom had grown up with the same patterns and thought they were normal. I told her I realized now that she had done the best she could, at the time. Needless to say, there was no response on the other end of the line. I thought maybe she had fainted and was out cold. So much for making peace.

In any case, she finally did respond, so I knew she was still conscious, and we talked a while longer. When I hung up I felt the weight of many years fall from my shoulders. With that encouragement, I proceeded to my ex-fiancee, Chris, and apologized for my rudeness the last time we had talked. He, too, responded in an understanding way and apologized for his behavior as well. So, two down and a lot more to go.

As I went through the list, I began to feel better and better about myself and lighter and lighter. When I was done I have to say I really liked myself. I liked the fact that I did not try to defend my actions or make excuses for my behavior. Instead, I apologized and that was that. I realized it was not about right or wrong at this point, it was about release and making peace.

By the way, *Level One* took a couple of weeks because I really had to wrestle with the idea of calling two of the people on my list, but in the end, my Higher Self won out. When I finished this list, I began another. Was I masochistic or what? It seemed I was on a roll, but when it was done, I knew it. There were no more people to release. I felt much lighter.

During this time I went through a liver cleanse which gave me incredible energy. I felt great and I was so energized that I almost couldn't stand myself. It was like having layers and layers of weight peeled away. I also had numerous past-life recollections during this time that dealt with anger issues. These were also released through the liver cleanse.

I began having recurrent dreams that left me angry. It seems I went around daily, exuding anger. It was a really strange time. Now I know how people who carry mountains of anger inside feel. And other people pick up on it because it's in their energy field.

I needed to release this *stuff* but as soon as I released one block of it, another would surface. Not only was I releasing my stash of anger from the last three years, but Joscelyn's anger also. And it wasn't only from this lifetime but from previous lifetimes as well. This accounts for the increase in past-life recall during this period.

I noticed I kept drawing people back into my life with whom I had issues. They would usually appear after I had released them. The most wonderful thing happened—I noticed I didn't feel the old feelings of anger toward them that I normally would have. I came to realize later that this happened to give me confirmation that I had, indeed, completely released them.

And one other thing: Not only did I not feel anger, I actually felt compassion for them. What an amazing thing! I really liked myself then and, liking myself, I came to realize, was the beginning of self-love.

Level Two: Managing Anger

Level Two involves learning how to manage anger and not store it. In this level you learn how to confront quickly and with compassion. This level is challenging for just about everyone. It was for me. In this level, pain becomes the impetus for confronting quickly, because you will actually feel the anger in your body if you don't confront it to clear the issue.

Any time we feel anger and don't release it, we are storing it in our bodies. I might add that the pain can be quite severe and can totally ruin your day, so confront quickly whenever you have issues with someone.

Remember, anger is emotional poison, it is toxic to your emotional and physical bodies. That is why you feel it as pain. Once you have cleared the issue you may find that you need a massage or energy work to heal the area where the pain was felt. This will remove any residue and keep your energy flowing smoothly.

During this period you may find that you draw people into your lives who will challenge you in many ways. Believe me, this is an opportunity in disguise. It gives you numerous chances to practice effective confrontation, and, I promise you, this level will not be considered complete until you have proven to yourself, the Council, your genetics engineer, and your guides that you are at least passingly proficient in this area.

You may go on to other levels, but you will continue to work on this one until you get comfortable with it. Just remember, it's for your own good. Every time you go through a confrontation, it *is* for your own good. I had to keep telling myself that!

The benefit of *Level Two* is the ability to manage negative emotions, especially anger, so that you don't stuff and store it in your body anymore. This can be accomplished by using the Formula as soon as the situation or issue occurs.

My Experience in Level Two

This was a long level for me, lasting from early March to the end of the month. I suppose the guides thought I needed practice in confronting others. I got my best opportunity with my friend, Bobbie, regarding the pace I had set for my spiritual evolution.

As I said back in Part 1, Bobbie did not understand what was happening to me, and, therefore, chose to resist or hold back until she could make sense of it. Her motive, I believe, was to try to help me as she had always done, but this time she could not.

She was afraid my body would burn out. I felt constrained just thinking about slowing down, and would respond with "I don't want to slow down, I want to keep going!" I became angry and impatient with her. Of course, I had no idea where I was going at the time, but that didn't matter. I knew I was headed somewhere and trusted in the guides to lead me. This did not make Bobbie very comfortable. So, I called her and set a time to get together.

We talked and she aired her concerns. I explained to her that I really did not feel any need to slow down; that I was following through on a commitment I had made but couldn't remember. I validated her feelings and sincerely thanked her for her love and concern. I also agreed to honor my body more. Until I confronted her, I was in pain, with the pain increasing the more I thought about the problem and the longer I waited.

It took me only a few days from the time it started until I called her to confront the issue. God only knows how I would have been feeling had I waited a week or two more! In the end, we cleared the air by validating each other's viewpoints and reestablished our friendship.

Looking back, I see that the confrontations are better done in person than by letter or phone. It gives us a greater sense of release, and as far as the guides go, a better opportunity to learn and practice confrontation.

There was also another confrontation I had to do during this time, and I regret to say that I didn't do so well on this one. I wound up in a lot of pain and I had to go back and do it over again. It was with a member of one of my networks. Try as I might, I couldn't complete this one without anger.

It was done over the phone and I could have done better had I done it in person. I find that when I confront face-to-face, my ability to confront with compassion is much greater, and that is the other lesson of Level Two. You must not only confront quickly, you must do it with compassion instead of anger. Anger keeps the cycle going and nothing good is accomplished.

So I urge you to do confrontations in person and with compassion. When I had confronted Bobbie correctly, my sojourn in *Level Two* was complete. The guides obviously felt that I had learned this lesson and moved me on to *Level Three*. Of course I didn't know what *Level Three* was at the time.

Level Three: Clairaudience

In Level Three your channel begins to open more, and you begin to hear your guides speaking to you. This is also the time when your light body begins going to the ship for the removal of your implants, and a Caretaker is left on your behalf to continue in your stead.

Caretakers are beings who have not been in physical form before. I am told they are young souls who have signed up for this project as a way to further their own spiritual growth, through care-taking or "house-sitting" for the bodies of those going through recoding. They need to learn how to function in our dimension.

Part of my contract in recoding had to do with giving two caretakers, Rachel and Shashara, the opportunity to train, using my body. This began in *Level Three* and was complete by *Level Eight*. By that time Rachel had gone on to be a caretaker for my friend, Pat.

We do not begin *Level Three* until the guides feel that we have learned and are capable of managing anger. Anger is the biggest impediment to recoding and can create setbacks in the process. Therefore, it may take a while for some people to finish *Levels One* and *Two*.

The benefits of *Level Three* are the ability to begin hearing multi-dimensionally and the beginning of your implant-removal process. Your ability to hear comes from the increase in your frequency. Your heightened frequency comes from using the Formula to release and clear the emotional blocks from your emotional body.

Your genetics engineer, who is assigned to you at the beginning of this level, is able to start removing the deactivated portions of your implants because you have cleared layers of your blocks. Your blocks are in your emotional body, and your implants are in your astral or light body.

As you release a layer off a block, your genetics engineer is able to then remove that piece of the implant in your astral body that corresponds to the block in your emotional body. Your genetics engineer is simply removing the portions of your implants that you have deactivated by using the Formula to release the layer on the block.

My Experience in Level Three

I began this level on March 31st. I know this because my journal has an entry about a dream within a dream that I had the night before. I dreamed someone was touching me. There was a man poking my leg under the covers with his hand. He was talking to someone else as he was doing this—the way a doctor talks to his nurse while examining a patient. I tried to open my eyes but couldn't. No matter how hard I tried, I couldn't wake up. It was that drugged feeling you experience when you wake up in your own dream, yet it wasn't a dream. It was really happening.

There were people jumping on the bed and talking to each other. They wrapped me in my covers and began lifting me out of bed to take me somewhere. At this point I became frightened and said "No!" That woke me up. Then I realized it had been the guides and Joysia, and I felt ashamed that I had been so frightened. They were taking me to the ship to begin working on my implants, and I guess they didn't realize I could wake up.

At that point they must have done something to block my ability to wake up because I didn't remember anything after that or after any of the subsequent trips, until I reached *Level Seven*. You see, they learn from us as much as we learn from them. Joysia learned he had to block my ability to wake up during the trips to and from the ship. I wonder if my experience put a scare into him. Ha! Ha!

Of my two caretakers, Rachel was in charge because she is a little more experienced. Shashara was a younger soul and totally inexperienced. She could care- take only when Rachel was with her. She could not come alone. When they first began coming, I didn't know why they were here. It was as if one day I was here, and the next I was gone and someone else was using my body. As the days went on, I was gone more and more. I really liked leaving, but coming back was a problem.

I remember being gone once for three days, and when I returned nothing had been accomplished. Marla, my business partner, said I had not been myself and that I was like a zombie. This is when I realized something was amiss. I had a discussion with Devin and Joysia, and they explained what was happening to me.

They explained, for the first time, the role of the caretakers and said I was going away to a parallel lifetime. Devin and Joysia always allowed me to experience the level first, before explaining it to me. I had agreed to give the caretakers use of my body for their initial training. But they also said I was to help train them, and one way was to teach them to be productive while I was gone.

My understanding of the caretakers came after a couple of strange and bewildering circumstances between me and my two friends Pat and Bobby. I had gone to Pat's house and, while there, she began asking me questions regarding a particular situation she was in. I did not answer her, someone else did. Prior to this, the day before, I had been in the office talking to Marla and someone else began speaking with her. It wasn't me. It was Roland, the one who got me hooked on "Jelaila" cookie dough.

Roland, my new guide, presented himself to me as a 16th-century shipping magnate, who also owned a string of taverns. Roland is a portly gentleman with a Falstaffian personality. He showed me a scene in which he was standing at the bar in one of his taverns, drinking and joking with the patrons, all the while mentally tabulating the amount of money he was making off the beers. Yes, he was a man with quick wit and a razor-sharp mind.

Roland said he had come to experience 3rd dimension again, and to assist Marla with the business while I was gone. One minute it was me talking and another I was gone and someone else was talking in my place. The shifting of personalities should have been disconcerting, to say the least, yet it didn't bother me at all. Not only did I have caretakers shifting in and out, but also my guides were getting in on the action as well!

A few days later, I went back to Pat's house to channel for her. I began channeling her guide, but before long Rachel and Shashara came through. They asked Pat if she would mind if they opened my eyes so they could see into our dimension. Pat said they looked around the room and just stared at different things. They would ask questions about the different objects and she would answer. She said she felt comfortable with them and had a blast helping Rachel and Shashara. Then I came back and she told me what happened. I told her I didn't remember where I had gone, but I knew that I had liked it.

After this Pat and I went to Bobbie's house, which is only a few blocks away. I could get there with my eyes closed, because I had driven it so many times. But not this time. I got lost and arrived 15 minutes later than I normally would.

I quickly realized the shift had occurred and someone else had driven my car! They must have had a lot of assistance and protection to get me to Bobbie's house in one piece, but the one thing they didn't do was access my memory banks so they would know how to get there. This was another indication that it wasn't me, but someone else.

At Bobbie's house they did even more shifting, and this made her very uncomfortable. She was trying to carry on a conversation with me and ended up talking to two or three different people. She was quite put out. Bobbie, just like anyone else, likes to know to whom she is speaking.

In the end, I was told, one of them spoke at length with her to assure her this was all part of the plan and everything was happening as it should. This didn't help much since Bobbie and I were not privy to the same information that the Council was.

She suggested I call Kim, who works with walk-ins, to see what was going on. Bobbie said this was like being possessed by two or three different people who could take over my body at will. I had felt fine with it. By the time I left her house, I felt dejected, rejected, and depressed.

So I called Kim, in Fort Collins, the following day, and she said I had made a contract with these individuals to allow them to use my body to experience our dimension. She gave me the names Rachel and Shashara and confirmed the name Roland. I had known about Roland before and knew he wasn't a caretaker but that he was a member of the Council, in charge of business affairs for Marla and myself when both of us were gone. Marla was also recoding at the time.

Kim stated that many people would be uncomfortable with this arrangement because it sounded a lot like possession, but it wasn't. Later, in talking with my guides and the Council, I learned the whole story about the caretakers. A rather interesting roundabout way, don't you think?

So here I was with two other people sharing my body and me, flying off to visit parallel lifetimes or aboard ship for recoding. It was like an extended vacation for me. Although some of the time away was spent on the ship, a good majority of it was spent in the parallel lifetimes.

Later on I would find out that one of these lifetimes was as a hypnotherapist and psychologist in California. No wonder I have always been extremely interested and fascinated with the mind. I have within Joscelyn's memory tapes her desire to major in psychology, and when I walked in, my first major accomplishment was to write a training manual which included a large section on relationship training. Yet, I had no prior training in this area. Neither Joscelyn nor I had ever gone to college, at least not in this lifetime. We were both self-taught.

I understand that most other people won't spend much time visiting parallel lifetimes. I went through this because the caretakers needed time to use my body for their training. Everyone else will not be gone that much and when they are gone, they will mostly be aboard ship for recoding.

The time away got to be like an addiction. I was gone more than I was here, only I didn't consciously remember where I went. I only remember the feeling of liking it there and wanting to go back. It seemed I was bringing back knowledge, which I was storing for later use. The time came when I had to make a choice to either stay and complete my mission or leave. I had become more attached to leaving than staying in this, my present reality.

This decision didn't come until I had finished *Level Seven*, several months later. It must have been trying for my friends to carry on our friendship with two other people present. I love them for sticking with me, even though none of us understood what was happening at the time. Most friendships would have been destroyed, or at least permanently damaged.

Changes came rapidly, and change is hard to take for many people; not to mention change of this nature. I had to constantly remind myself to remain open to what was happening, since there was nothing in my collective conscious past with which to compare it. I went on faith that I was being guided by loving people who knew what they were doing.

After this interesting introduction to *Level Three*, Joysia told me to remove myself as much as possible from electricity, as this would interfere with my new, expanding channel. I was becoming clairaudient. The electricity, he said, creates static on my channel and is felt as physical pain. (I had to agree!) He said I would only have to make this adjustment for a short time. This message came home to me in a rather uncomfortable lesson.

I had taken my daughter and her friend to one of those indoor playgrounds for kids—you know, the kind where the noise level is so high you wish you had remembered to bring ear plugs. The next morning I woke up with pain coursing through my body and an unbelievable headache over my right ear (I hear clairaudiently through my right ear).

When I checked in with Joysia, he very diplomatically reminded me about the statement he had made to me only two days earlier about staying away from electricity. Needless to say I didn't have to be told again.

You may find that they will urge you to go to places that have no overhead power lines, so they can speak with you. One reason they do this is to be able to check your channel and then make adjustments the next time you're aboard ship. The other is to give you an opportunity to speak with your guides through your newly expanded channel. I remember it as a very heart-warming, loving, and emotional experience.

It was also in *Level Three* that I began channeling the material for this book. I was told about it during a walk in the woods. There is a special park, named Shawnee Mission Park that is near my home. My guides send me there when they need to have a long conversation with me. They know I can concentrate best when I am there.

It was April 29th, and off I went to the park. Actually, I could have begun receiving the book sooner but I blocked it because I didn't have a computer at home to work on. The CRN office was also out of the question with the constant interruptions of the phone and clients. Well, the guides took care of that.

I have a channeling client who graciously offered to loan me a 486 computer he wasn't using and as soon as it was set up on my kitchen table the book was begun. I would type for hours at a time. On one Tuesday I typed from 8:00 a.m. to midnight with only three short breaks. The guides were sending the information as fast as I could type it.

During this period I began receiving more and more information on a variety of topics. It was similar to a flood gate being opened to the other side. I was conversing with my guides almost continuously. My method of talking with them had changed. I used to go through this long procedure of protection, light-trance meditation and relaxation before being able to hear them. Now I could just tune in, listen and hear them talking. It was sort of like tuning in to a radio station and turning it on and off at will. A lot of change and growth took place during this level. It was unforgettable.

Level Four: Clairvoyance

This was a wonderful level for me. In *Level Four* your third eye begins to open. You first begin to see darker energy forms, so don't get frightened. These are usually negative entities and you can see them more easily because they are denser. It's a good time to learn how to work with these entities. They, too, are souls and part of Divine Creator—they are just playing a dark role. They too are learning. You usually see them attached to others. Bless them and send them on their way but, know they will most likely return to their host, unless that person requests they leave. This will give you a good opportunity to work on any fears you may have about negative entities. They cannot harm you unless you allow it. Remember, you are in control here.

After a while, and it depends on the person, you begin to see lighter forms and then colors. Some people think it happens all at once— one minute you see nothing and the next you see everything. This is not so. Remember, your third eye has atrophied from nonuse. It's like a muscle, it must be exercised to make it stronger. So practice, practice, practice. Get a friend who has just had someone who recently passed into spirit to work with you. Make sure you know nothing about the deceased person. Concentrate on them and see what you get. Don't be discouraged if nothing happens the first time. After a while it will happen. You will begin to see and hear them. Channeling or trying to see each other's guides is another good exercise.

Some people experience seeing blue lights and twinkling lights. This is all part of it. Clairvoyance is an ability that must be honed. You can also practice with your guides. Ask them to present themselves to you in your mind and muscle check to see if you are describing them correctly. This will quickly build your confidence in how well you are seeing.

Once again, the benefits of *Level Four* come from raising your frequency by using the Formula to release layers from your emotional blocks.

My Experience in Level Four

Shortly before beginning this level I was urged by Devin to go to the bookstore for a book. This is a normal way Devin and the others communicate with me. So here I am at the bookstore, once again, not knowing what I'm here to get. I do my usual thing and browse until one catches my eye. I've actually had them fall out and nearly hit my foot. I guess this is how Devin makes sure I pick a certain book, and usually they have been some of the most important ones for me.

This time the book was *The Eagle and the Rose* by Rosemary Atlea. It's about a medium who is clairvoyant and her initial struggle to accept her gift, which she had been born with, and to use it to benefit others. I didn't start reading it until almost a week after I got it, which is highly unusual for me. I normally start them right away.

About halfway through the book I realized what *Level Four* was. It was about becoming clairvoyant. When talking to the Council that night, I was told that this was another way they would show me what was just ahead in the recoding process. They would impulse me to go purchase a book that covered the next level. I must say I like this form of guidance.

During this period I began seeing darker energies. The first one I saw was attached to my friend's boyfriend. It was a big one too! My seeing it gave her confirmation that it was there, as she had already felt its presence. She had had another one of our friends look at it and this friend confirmed it, too.

After that I began seeing lighter energies. I saw colored lights and geometric forms. Then I saw my first person, but he wasn't dead. He was still alive! He was John, the ex-husband of my friend, Pat. She had come for a session with her guides. I was beginning to channel regularly by this time.

He was sitting in a chair (in spirit) across from her in evening clothes with his legs crossed. John told me he had come to give her a message. When I gave it to her she began to cry. She said the message was something that had driven them apart, and now he was beginning to work on it. What confirmation for me!

During this time I was becoming more and more unhappy with CRN. I wanted to channel full time and write this book and CRN was keeping me from doing it. I knew in my heart that I had to make the break, I just didn't know when. Reading the book by Rosemary Altea had triggered something in me.

I wanted more and more to channel full-time and hone my budding psychic abilities. At that point I made the conscious intent to do just that. It wouldn't be long before that intent became my reality.

Rachel was getting used to using my body, and we were kind of settling in together. I was going away more and more and leaving her in charge. There was only one problem. She did not have the emotional connection to my life and this dimension that I had, so she came off as being unemotional.

This left those interacting with me feeling a little cheated and angry. It left me (Rachel) feeling depressed. I would come back and there, stored in my memory tapes were days of darkness and sadness. I talked to Rachel about this and she said she would work on adjusting her vibration to mine so she could access my heart energy.

Little did I know Rachel was having trouble because my heart chakra had never been fully opened to this dimension and she couldn't open it without my permission. Neither Joscelyn nor I realized it was closed. Opening this chakra would come later. I also found out that others had the same problem, which indicated that their hearts weren't open all the way either.

As a result, the caretakers could only express love to the degree that the heart chakra was open in the person's body they were caretaking. I began this level on April 30th and finished it on May 11th.

Level Five: Integration and Contract Alignment

This level involves the integration of the first four levels. It is similar to a clean-up of unfinished pieces or a tying-up of loose ends. This doesn't mean everything in the first four levels will be completed. You may find yourself going back and working on some of these levels when you are in one of the last four levels. This just means that what can be done and completed is done at this time. This is the first part of *Level Five.*

The other part of this level involves aligning you with your contract. What this means is, if you are not fulfilling your contract through your present line of work, you will be moved to a more fulfilling area of work before completing this level and finishing recoding.

This can be very frightening for many people as they will see doors of opportunity closing and sources of income drying up. During these times it will be easier if you remember you aren't going to starve (what good purpose would that serve?), and that the universe will support you. This will hopefully make it easier for you. It did for me.

Also, remember that until you have stated to the Sirian/Pleiadian Council of the Galactic Federation that you do, indeed, intend to honor your contract, though you do not remember what is in it, you will not be able to complete this level. Honoring your contract without knowing what it is, is a leap of faith, or, as some people say, a jump into the void. You may find this much easier to do when you remember that it was you who created the contract and you wouldn't be honoring something you weren't capable of doing.

The benefits of *Level Five* are that you gain clear understanding of your contract for service to humanity and a sense of purpose to fulfill your contract. Once again, this can only be acquired by releasing enough layers from your blocks to raise your frequency to the level that allows you to access the information on your contract.

My Experience in Level Five

This level was a real challenge, but not the most challenging of the nine. This was when I came face to face with my need to leave CRN. I knew I wanted to but I couldn't make the jump because I was afraid I wouldn't have the income to meet my monthly financial obligations. Sound familiar? I would have to start over from scratch, with no reserve income and only a few clients.

To make ends meet I would have to have ten clients a week. I was averaging two a week at this point—a big gap! Well, the universe had a way of nudging me toward the appropriate move. All my options dried up. I had a lot of new opportunities starting to cook with CRN, and I stood by helplessly as each door closed in my face.

Now, as if this weren't enough, my partner Marla, had new opportunities and doors opening for her, right and left, and we were in the same company! I kept asking, through my tears, what's wrong with this picture?

The straw that broke the camel's back happened after a particularly trying week. I was at home alone and crying (not unusual during this time), "blessing" out my guides and anyone hanging around up there in my vicinity. I kept thinking I needed to call my friend Pat but thought better of it, because I didn't want to burden her with my problem. As it happened, she called me. Interesting how these things happen, isn't it? So we talked and I poured my heart out to her. I said, "I'm tired of CRN, I'm sick of the way my life is going, nothing is working right, all my opportunities are drying up but Marla is doing great!" Then I cried.

Pat was silent for a moment and then she said to me, "Jelaila, did you ever consider that you need to be channeling for a living?" At that moment I knew why she had called me. I also realized what *Level Five* was about. I was finally aligning with my contract which was to be a full-time channel. That wasn't all of it, but it was to be my primary job—a channel for guides and ETs.

At that point the clouds of depression scurried away and the sky in my world turned a beautiful blue. This was June 28th. On August 1st

I officially ended my role as founder and president of Creative Referral Networks, Inc.

In the meantime I was chomping at the bit to leave, and the money fears didn't faze me anymore. I also began getting calls for channeling. Things began to take off—doors were finally opening! This was a far cry from the attitude I had had when I entered this level. Gone was the fear of making ends meet. In its place was a burning desire to pursue a line of work that made my heart sing. *Level Five* changed my life!

Level Six: Taking on Your Power

This level differs in intensity, depending on how much you are able to own your power and how much you give it away. I have heard others talk about their experiences in this level and they all vary. *Usually you draw situations into your life to make you aware of how and where you give your power away.* For some it is a person who challenges them in some way—for many, it is a person who is a sexual partner and love interest. In most cases there is anger involved, but do not judge the anger here as negative.

When the person challenges you, it is natural to react with anger. They are taking your power, but you are giving it to them. Use the anger to identify your feelings. Let it be an indicator to you of how you are giving your power away and getting angry about it.

An example could be an ex-spouse whom you feel allows your children to control him or her. The children are not getting disciplined. Another example is a co-worker who takes credit for your ideas and work. Any situation that brings up your anger is an indicator of where you give your power away.

Be aware, and keep a journal during this time. In fact, it is good to keep a journal of the whole recoding process. Once you have completed it you can look back and be amazed at how different and more powerful you are now.

Once the Council and your guides are as certain as anyone can be that you will use your power for the fulfillment of your contract, the lessons will begin. Understand that you will *be challenged only in the areas where you tend to give your power away.* This challenging acts much like the pain in *Levels One* and *Two.* It is an indicator of where you need to grow. I have been told by the Council that only those individuals who agree to use their power wisely and for the fulfillment of their contracts will complete this level.

Once you begin taking on your power, your ability to manifest your thoughts is increased. The old saying *beware of what you think because you will get it* is quite appropriate here. Keep positive, uplifting thoughts in your consciousness as much as possible. Any negative thoughts about yourself or others will manifest and boomerang on you rather quickly. Remember, everything cycles in the universe, so your thoughts will come back to you—only now they will have much more power!

The benefit of Level Six is your ability to take back your power from those you have given it to. The ability to hold on to it and use it is another benefit. This is accomplished by releasing the layers of your blocks that contained lessons of powerlessness. The ability to hold and use your power will increase the more you integrate the lesson of powerlessness. This one usually involves many lifetimes and many layers, so don't get frustrated if it takes a while.

My Experience in Level Six

I began *Level Six* on a high. I had just accepted my contract to be a channel for other peoples' guides, angels and ETs. I was in a place of calm and serenity, but this was not to last for long. My time of peace came to an abrupt end on June 30th.

This was the first group channeling session for recoding that we did. My friend, Ann, and I had decided earlier in the month that we needed to start a group session for those interested in accelerated recoding. We would hold it at Ann's house until we outgrew it. I would channel Devin/Anu and Joysia, and the group could ask questions. We made a list and started calling. There were a few who had already begun the process, and they put out the word as well.

Twenty-five people showed up for the first session. This was a good turn-out since we were instructed by the Council to only invite those we felt were ready for the process. We were not to send flyers or advertising to the local metaphysical papers or bookstores to promote the event. That meant those who attended would have to be pretty knowledgeable about metaphysics and would have done a lot of clearing. We did not feel we were the ones to decide who was ready, so we asked the Council, our guides and the guides of those who were to be there, to help us. The accuracy rate was quite high.

It was a Sunday night, and the moon was full. What a perfect time to launch this worldwide project! It was also near the Summer Solstice. Had I known what the Council had in store for me, I most likely would not have shown up.

After the introduction was completed, I went into trance. The energy in the room was terrific. There was a ship stationed over the house, and it was beaming energy into the room—a great setting. When I came out, the energy had changed. It was not positive, loving, and uplifting as before.

I could feel tension, and I realized I felt as if I had been hit in my solar plexus with a big blast of anger. It seemed Devin/Anu had pushed the buttons of most of the people in the room. Devin had channeled to Anu, who in turn channeled to me the information Devin wanted shared with the group.

Devin and the Council had chosen not to make a distinction between the Earth Sirians who needed the new bodies created by Enki and Ninhursag, and the Walk-ins and Starseeds who weren't of the same soul group. Most of those in the room were walk-ins and starseeds.

He also did not clarify the black race as the indigenous race who would later, after a cross with the Nibiruans, become the land guardians of Earth. And he called everyone lulus. This meant primitive worker in his language, but this audience didn't know that and you know what we mean when we call someone a lulu—it's not a very complimentary term in our language.

Needless to say, Devin/Anu said a lot of things without clarifying them or putting them into proper context. It was as if he had staged the whole thing. Many people left feeling angry and insulted. They also questioned his ethics and intentions. One person even asked if he was of the Light! My friend Bobbie left in the middle of the channeling. This put the icing on the cake. People looked up to Bobbie and when she left, it made them question the integrity of the Council and myself for channeling them—a rather unpleasant outcome to an eagerly anticipated event.

Even my friend Pat was angry with Devin and Anu and said she would not attend again. Not everyone was upset, only some of them. The point is, no one should have left upset. It should have been an informative and uplifting experience. Instead, Devin/Anu came off as arrogant and emotionally uninvolved.

The only thing that saved the session was Joysia. He is not of the Council but he is a consultant to them and head of the genetic engineers. His talk on recoding was most informative. No one seemed to have a problem with him. Thank goodness Joysia talked last!

When it was time for the session to end and for me to come back, I wouldn't come back. Joysia had to ask for assistance from the group to get me to come back. I only remember feeling a heavy weight in my power chakra and saying to my friends in spirit that something was wrong, and I didn't want to face it. I wanted to stay with them. I even began arguing with them, but they insisted I had to return.

When I regained consciousness, there were five people hovered around me, and Marla was holding my hands. I started to cry and couldn't speak. I was so sad that I had to return. I didn't want to come back, but I didn't understand why. It would only be a few minutes before I knew.

This, of course, upset the group further. Not only did Devin/ Anu act up, but the channel didn't want to return. Not a fun time at all!

It took two weeks to clear this up. At first, no one talked to me about it. Pat would not talk to me and neither would Bobbie. This was very upsetting. The only one who did talk to me was Ann, and she saw it a little differently than the others, maybe because she had had prior conversations with Devin and knew what he was really like. Yet, she was puzzled by his behavior too.

I could not understand what could have happened to cause this much discomfort between me and my friends, so I called them. First I called Pat and heard her out. Later in the week I called Bobbie and we arranged a time to get together so I could hear her out. I listened to the tape and I must admit Devin could have done a better job. What I could not understand is why this happened. Why would he do this to me? I went through days of yelling at him and the Council. At one point I was ready to fire him.

Bobbie *really* didn't like him now, but I knew it was due to a past-life situation between the three of us that she needed to resolve. She felt he was controlling me and not acting in my best interest. Based on Devin's and Anu's performance I could easily understand how she could reach this conclusion.

During this two to three week period, I went through cycles of depression and even lost faith in the Council and the spirit world as a whole. I was like a ship that lost its rudder and was left to drift, lost and alone. I even lost faith in myself and my contract. I thought there must be a mistake here. I began to think that I had agreed to assist the Council only to be used by them. So I decided to fire Devin and the entire Nibiruan Council and go get a 3D job!

The turning point came on a Friday when I was at my lowest point. Bobby urged me to go above the Council to seek help. I did. I went to Sananda. I visited with him, and he told me I did not have to work with Devin if I did not want to. He gently reminded me that I had come here to serve and I had to look within my heart to determine if what I thought was my service, really was. This was the turning point. Over the course of the next three days, I came to realize that DNA Recoding was my service, as well as channeling.

The most startling revelation was that I was the one in charge down here, the Chief Emissary. I began to understand the part the Council played and the part I was to play. My contract was to lead the way on Earth and with the knowledge I brought, empower others to be leaders. That was why I walked in. Instead, I was waiting for them to tell me every move to make instead of me directing the traffic. This was why there were so many setbacks, they were waiting for me!

Now I understood why it had been so hard for me since I walked in. I was expecting them to provide for me and manifest for me. They wanted me to complete the RRA process, create the template, and then they would learn from my experiences and do the rest. I needed to use my own power to take care of me, not to have them do it for me.

I finally took them off their pedestals, something Devin had been telling me to do for a long time. I thought I had, when I began to see him as a brother and former father and lover of many lifetimes. But I hadn't when it came to them doing for me.

Once I realized this, they began to communicate with me again. During the time from the channeling to that point, I had not heard from Devin, or anyone else on the Council. It was as if they had disappeared, which somewhat puzzled me at the time. But you can bet they heard from me. I was yelling at them almost daily. I wouldn't blame them if they had chosen to wear earplugs.

We made up, and I apologized to Devin for calling him names and yelling at him. He was a little reticent at first, but finally, gave in. This had been a very trying time for both Devin and me. We forget that ETs are simply people living in another dimension, and they have feelings, just like us. We expect them to be all-perfect and all-loving.

Devin told me he was very sad that he had to put me through that experience with the group, but the Council had agreed that it was the only way for me to see where I was giving away my power. Devin explained that I had not been monitoring the entities coming through me, and this was unacceptable. It put me in a position of powerlessness. I found trance channeling to be easier. He said that trance channelers can't monitor and, therefore, are in a compromising position, especially with the amount of dark energies arriving on the planet. He said he understood my anger toward him, and he didn't hold it against me. He really is a sweet, loving being.

In the end, I learned that they did not want my power. The Council had, indeed, acted in my best interest in getting me to own my power and not, as Bobbie had so accurately perceived, give it to them. Though it was unpleasant, the actions they took helped me to take back my power and perform the duties I agreed to do to help them.

I was here to run the show, and they were here to assist; not the other way around. I also learned it was not up to them to provide for me—I had the power to do it myself. Once this was learned I was ready for the next step, *Level Seven*. I learned a lot in this level. I was now able to hold my own power. Now I just had to learn to use it.

The benefit of *Level Six* is the return of your power. This comes about because you have released layers off the blocks that deal with powerlessness. Powerlessness and anger usually come together.

Level Seven: Using Your Power to Dissolve Fear

This level is a challenging one because here we learn to use our power to illuminate our fears so we can see they are only illusions. I liken it to navigating a long twisting tunnel in the dark. There is no light at the end of the tunnel until we realize that we must put it there. No one will put it there for us. As we move through the tunnel, we bump into obstacles (our fears), and we have to use our torch (power) to illuminate them to see them for what they really are.

Over and over, our fears will be presented to us through situations that arise in our lives during this level. For many, these are experienced through cycles of joy and then depression. Once we realize that our fears are just illusions and integrate them, we will be done with this level. Once completed, our remaining implants involving fear are removed in the next level.

The benefit of *Level Seven* is the accelerated removal of layers from your blocks. Though intensive, it provides as thorough a cleansing of your emotional body as you are able to handle at the time. This intensive removal process will heighten your frequency almost to the point of activating your MerKaBa. Another benefit is the increased feelings of togetherness and power you feel because you have had to use your power to see through the illusions of your fears. At this point you begin to handle fear much more easily and effectively when it arises, no matter how it manifests.

My Experience in Level Seven

I thought the worst was over when this level began. After figuring out that Devin had staged the whole scenario at the group channeling session and then understanding and integrating the lessons involved, I thought I had come through the most difficult level of the nine. But then came *Level Seven*. This was not as socially challenging as *Level Six*, but it was emotionally challenging.

I found myself going through three-day cycles of joy and then depression. My fears came to haunt me in the form of people and situations. I remember three distinct cycles during this period where I came to terms with three different and powerful fears.

Now, being a recent walk-in I didn't have as much fear as someone who had been here on Earth, basking in the frequency of fear all their life, but I did have my share. These were around manifesting money, speaking the truth about my contract and my history in world affairs, and opening my heart to this dimension.

This first one, money, was the most challenging. I guess that's why I must have agreed to tackle it first. After the disaster at the first group channeling session, my private channeling business dried up. The bills were coming due and no money was coming in. It reached crisis stage one day, and I sat down and cried.

It was then that I understood what was happening. I had brought this forward so I could see that my fear of not having enough money was all an illusion. So I used my power by stating to the universe that I was no longer buying into this game. I said "I know that I can manifest whatever I want. I just have to do it."

Needless to say, my business started to improve. The phone began ringing and I earned more money in the next week than I had the entire previous month! I visualized money as energy that was being blocked by a window. So, in my imagination, I simply opened the window all the way and "saw" the money energy flowing through, uninhibited.

The next illusion of fear was taken care of by my conversations with Sananda. He helped me to see that the world didn't have to agree with everything I said, and that I was to speak the truth for those who had ears to hear. He explained that many a prophet and wise man or woman had been ridiculed for speaking out.

Most people who incarnate here on Earth are here to learn lessons through the medium of fear. He said there had always been someone sent to Earth to be a beacon of light and to remind people of their true identity. And they knew ahead of time that they would not be welcomed and that they would only be heard by a few. He said it was for the few that they were sent, just as now.

Not everyone, he explained, is supposed to hear and accept, only those for whom I was sent. He told me not to worry about the rest. With Sananda's words I was able to move through and banish this fear, only to find residue still around later that would inhibit me from writing on the book for two more months. This was later identified in a channeling session with Ann, Devin/Anu and Joysia. Then the residue, too, was released.

The final illusion was the one that was the easiest to see through. I was assisted in this one by Ann. We had been mother and daughter in a previous life. What would I have done without her during this time? She brought it to my attention one day, when I was at her house to edit some recoding material I had written for a metaphysical magazine. Ann said she felt we were not as close as a mother and daughter should be because my heart was not open to this dimension. She was aware of this because she had recently become aware of it in herself, and she proceeded to help me open my heart.

Bobbie had also assisted me here. I was at her house one afternoon, and as we were sitting outside on her patio, she made the same observation. She also said that if I were going to be here I might as well enjoy it, and that this block could be the reason why I had not manifested the kind of life I wanted, since I had walked in.

Ann helped me to open up through speaking to the archangels involved, and I worked on my image of what I wanted life to be. In the end, I banished my fear of Earth because I realized that enjoying it would not make me a prisoner here and that I could still go home after my mission was complete.

She used soul clearing and holographic repatterning. When she repatterned my closed heart chakra I let out a bone-chilling scream and with it went my guilt and shame over the terrible way the Lulus were treated by some of the less-evolved Nibiruans. Of course I am referring to my 6th-dimensional aspect, Ninhursag. I had held on to this pain and shame and understand this to be the reason I am adamantly opposed to doctors and why I am a healer myself. With this under my belt I began Level Eight.

Level Eight: Releasing the Last Fear Implant

Level Eight was much easier than the others for me. It appears that the most challenging levels are the first seven, and the last two are the final clean-up.

Any lessons around fear that need further work will be completed here. This usually comes in the form of confrontation with someone or some situation that makes us fearful. Otherwise this is a rather pleasant level.

Many people begin to experience dreams about flying. This is to get us in touch with our new reality of no fear. In this new reality we are not as heavy because we have cleared so much fear-laden density, so, now, we can fly. These are usually very exciting and uplifting dreams.

Level Eight is also where the last implants involving fear are removed. By the end of this level we know what it feels like to live in a world of fear but not be affected by it. This does not mean we don't experience fear as a natural reaction of our conscious mind. It means that when this occurs we are easily able to see what is happening and let it go.

The benefit of *Level Eight* is the feeling of even greater lightness and the feeling that you could fly. You are very close to activating your MerKaBa when you complete this level. Once again, use the Formula to move quickly through the test that is inherent in this level. I did not have it when I went through this level. That's why it took so long and was so emotionally challenging. As soon as the

My Experience in Level Eight

This was a wonderful level for me. It began with my being able to manifest the perfect roommate, which had eluded me in *Level Seven*. Now that I had one, I didn't need one. I also saw that people were attracted to me and wanted to hear what I had to say. The group channeling sessions were proof of this. They had made a complete turn-around, and the number of attendees were growing each month. Devin and Joysia were doing a terrific job of explaining how we are all in this together and the job could not be completed without a team effort on the Council's part and on our part.

I was learning to direct my will to manifest what I wanted. This included making a plan. The guides who had done our manifesting for us until now had always created action plans. I was learning how to create good ones by myself. This first opportunity came about through none other than my soulmate and ex-fiance, Chris.

I said I wanted to understand why we had ended our relationship and why he had jumped into another one only a month later. I wanted to put closure around it. The plan included seeing him again and finding compassion for him by understanding the why's of his actions.

The plan manifested almost instantly. Chris called me, out of the blue, stating that he had some pictures of Danielle and our only Christmas together, and he would like to bring them by if I wanted them. Now, understand that I had not seen him since the previous October, eight months earlier. He showed up at my house one

Wednesday evening, and we talked for about three hours. This was probably one of the most honest and open-hearted conversations we had ever had.

I came to understand that Chris had jumped into a relationship with this woman because he was afraid of being alone. He said he realized that he had settled for less than what he wanted. He believed that one could not have both peace and passion in the same relationship, so he settled for the peace.

His life was stable but rather dull. I saw a man who had gone to sleep. There was no life in him as there used to be. I saw what fear could do. I finally felt compassion. I finally felt released. I also admitted to myself that I had asked for this girl to enter his life. I had begged the guides to send him someone who could love him the way he wanted to be loved, since I could not buy into his control game.

They did, and she was everything I wanted her to be. It's amazing what you can manifest when you want it for someone else. Now I just had to learn how to use that same passion to manifest for myself.

The next evening I was to experience another joy of *Level Eight*. I had gone to a local restaurant/nightclub to lend support to Marla and some other friends who were trying to promote a single's cruise that the club was co-sponsoring. I had not intended to be there long and had definitely not dressed to go into the nightclub. But when the group moved into the club from the restaurant I'm glad I decided to go with them instead of going home. I had just settled into a chair when a man came up to me and asked me to dance. Now, mind you, I usually don't get asked to dance. I am told men are intimidated by me.

Tonight was different. I was on cloud nine and feeling wonderful. It must be those men just wanted to bask in the glow of *Level Eight*, ha-ha! I danced almost every dance and one of the men even gave his card to Marla and asked her to tell me he wanted to take me to his resort in Jamaica. What a blast!

I understood, now, the power of attraction one develops once the fear implants are removed. People want to be near you because you are so light and your positive energy acts like a magnet drawing others to you. I had ten of my strands connected by this time. I could only imagine what it would be like to have all twelve connected!

Joysia informed me during this time what it would be like to have all 12 strands connected. He wanted me to understand that the fusion was being done in my astral body and the physical body was still too dense to handle that kind of energy. He informed me that the operations were done on my astral body and the effects were felt in my physical body just as everything else is. He next informed me that I had agreed to use my power to fulfill my contract and that everything I wanted would come through the fulfillment of it.

Joysia said he was concerned that many people believed they would be able to move mountains once their 12 strands were reconnected. He said that, yes, they could, but they would have to become adept in that area and that teachers would be provided to give that form of training. This would come in the form of people, books, experiences, and special guides.

He also informed me that my need of assistance by caretakers was almost at an end and that Rachel and Shashara would soon be going on to assist others. He said they were grateful for what they had learned through assisting me and they would use the information to train other caretakers.

I remember, in particular, one trip to the ship during this time. I woke up during the operation to find a quarter of my head had been taken off and they were working on a crystal located near my crown. The pain was intense, and they told me to go back to sleep, which I did. And I felt no pain.

I remembered looking around the operating room. I was lying on a silver table and there were instruments overhead on long cords. There were flashing colors around the room which I later came to understand were people. The atmosphere was warm and loving, and I didn't want to leave. It was an incredible experience.

When I woke up the next morning with no pain, I remembered the incident, and I knew it wasn't a dream. I had another episode on the ship in *Level Nine* which was quite different. I finished this level on a pleasant note on August 3rd. Somehow I knew that *Level Nine* would be just as pleasant. I was right.

Level Nine: Release of the Last Guilt Implant

In this level you will experience one last test. This test will be to see if you are ready to release guilt from your life and go on without it.

Many people use guilt as a crutch or excuse to keep them from moving forward.

This test will come like all the others. You will draw it into your life as a situation or person, and you will have to choose how you will respond and move through it. Once this test is completed, the implant is removed. You will also begin to notice more and more how you use your power to shape the outcome of events in your life. You will notice how you quickly work through conflicts and fear because fear is an ingrained response, and you will still feel it, but the difference is that you will be conscious of what you are doing and be able to release it.

You will see positive changes in your life and new opportunities coming to you, and now you will know how to take advantage of them. Things will begin speeding up, and balancing your life will be a challenge, but well worth the effort. People will be drawn to you and will want to spend time just being in your presence. They will feel your energy and want to bask in it.

This does not mean you won't have bad days. It means that when you do, you will know almost instantly what they are about. Your intuition is heightened now, and your abilities are being strengthened. Use the Formula to move through the guilt test inherent in this level.

My Experience in Level Nine

Level Nine started out quite peacefully—so peacefully, in fact, that I asked Devin why. I was used to being hit with big changes and this one came in like a lamb. But I was in for a pleasant surprise.

Ever since *Level Eight*, where I made my latest debut back into the world of singles and dating, I had been drawing men like crazy. They were coming out of the woodwork, so to speak. This was playing havoc with my time. I realized I had not made any time for playing, which was something I had said I wanted to learn while I was still in *Level Eight*. Joscelyn grew up in a family where work and responsibility were right up there with godliness, so she never learned to play. As a result, I am having to learn this, since there was no information left in her memory tapes for me to access.

So here I am, learning to play at almost 40. The guides said things would speed up for me, and they were right! I was out networking Sunday, Tuesday, and Wednesday nights every week and one Thursday night every month to build my channeling practice. My daughter came home on Friday night and stayed till Sunday afternoon. That left only Monday and three Thursday nights each month for me to play.

Exercise is a big part of my life so I had to make room for that too. I would go out to the park and walk six miles on the trails. I was doing this three nights a week. Where was I going to find time to play? Well, I did find time. I did my play after my networking functions, since they were over by 7:00 p.m. and I did my exercise during the day when I didn't have clients. I felt I had really tackled balancing everything when my guides brought to my attention, in a rather unpleasant way, that I had forgotten to make time for writing this book!

I had not worked on it in over a month. So I did some more juggling and gave up some sleep time. I like to take cat naps during the day, but used this time for writing. Once I had done this, my life seemed to smooth out, and everything fell into place.

So, balancing your time and life when things begin to speed up is a major part of *Level Nine*. It is one way you use your power. Many people give their power away by spending time rescuing others and fighting their battles for them. When you reach *Level Nine* you learn to hold your power and use it to serve others by fulfilling your contract. This does not allow time for co-dependent behavior such as rescuing others.

I also found I did not feel as scattered as I used to. I know where I'm going and what I am to do. When there is a glitch, like not giving time to the book, I seem to find the solution quickly—in this case through Ann. She did a holographic repatterning on me. I had a block around completing the book.

This block concerned Joscelyn's parents. It is my understanding that her mother, Modean, was Mogar in my lifetime as Hatshepsut.

Mogar was a high priestess and Joscelyn's father was a priest. They were responsible for my death, as they had been in league with Chris (Senemut) to have me killed for speaking out and exposing the priesthood for the scoundrels they were. I was concerned about their reaction to the book, as I am certain they will be exposed to it. It was a purely emotional response, brought forward from a past life. Ann helped me to clear it.

Another piece of *Level Nine* is choosing your vocation. I chose channeling, past-life regression, and what I call mediumship, where I am able to contact people who have recently passed over. I was especially interested in this area because I want to work with parents who have lost their children. Their grief is so intense and I feel that knowing their children are still alive, well, and happy would bring them great comfort. My guides went to work at once.

The first thing they did was send me to the bookstore to purchase books about mediumship and channeling. They told me they wanted me to read them so I would understand that I was already doing both and only needed to practice to hone these abilities. Also, finding out that other channelers experienced the same things as myself really helped me to see that I was on target.

I also wanted to have out-of-body experiences (OBEs), and go to the ship. After I remembered being there with Devin and Joysia, I wanted to learn how to visit at will. I was sent to pick up a book called *Soul Traveler* by Albert Taylor and learned, once again, that I had been doing this for some time during my catnaps. And, once again, I was told all I needed to do was practice.

The other thing I wanted to do was regress people into the past so they could bring their lessons forward and complete them. In my parallel lifetime in California where I am a hypnotherapist, I specialize in past life regression. I was told I need only tap into that lifetime to be able to regress people.

Now, although I had never done this before, it became clear to me that I had. When I used to go to Bobbie for regressions, she commented, more than once, how easily I seemed to tap into the Akashic records to experience past lives. I have also found it quite easy to regress myself, so I feel this correlates with the abilities for regressing that I have in my parallel lifetime.

On top of that, the guides also said that they would teach me how to regress people their way. The guides said they wanted to guide the client back and have their personal guides present to assist in choosing lifetimes that will bring the greatest awareness to meeting and solving the challenges the client faces today. All the guides needed was a good, clear, open channeler with knowledge and experience in regression work.

You see, guides choose people, based on the kind of information they have stored in their memory banks. Between me and my parallel life as a hypnotherapist, I had the right ticket. Now I regress people while both their guides and mine guide me. The results are profound, to say the least. I've even done an exorcism on a client!

I wanted to take a class from a leading hypnotherapist in town, but was encouraged by my guides not to. The guides said this would mess up their ability to train me their way. Right now, they said, I was like a blank canvas and they wanted to paint it their way. So I let them.

I am sharing this with you to help you understand that some of your training will come straight from your guides. What better training could you receive? I was able to be trained like this because I am clairaudient and my clairvoyant abilities are getting stronger by the day. So, as you can see, it seems we tend to choose vocations where we have innate abilities and all the guides do is help us perfect them.

I released my last guilt implant on the night of August 4th. I dreamed about it. In this dream I had a big wad of gum in my mouth and finally managed to spit it out. Next, I found myself needing to use the restroom so I moved to what I thought was a restroom and found it was, instead, an outhouse that was beyond filthy. On the back wall of the outhouse was a picture of a man and woman holding hands in a Victorian setting. This left me feeling sick and ill at ease.

It didn't take me long after I woke up to realize this was about releasing guilt, about speaking my truth, and that it had most recently come up in my (Joscelyn's) Victorian lifetime with Chris. I came to realize I carried guilt about speaking up to men.

Joscelyn's father in this lifetime was someone she feared greatly and, as a result, she carried a fear of speaking up to male authority figures. I inherited that trait when I arrived in her body. This was released after my proving in two separate situations, that I was ready. Let me briefly explain.

The first opportunity came through dating. I met a man who talked only of himself and his 90s opinions about equality and all that stuff. After two dates of listening to him expound on his virtues, I finally spoke up. This came about during one of his daily phone calls, which also irritated me.

I told him I thought his opinions very noble and such but said I didn't believe he walked his talk, as I had not seen any evidence of it. I also told him I was not interested in seeing him again, since the conversation was mostly one-sided and I wasn't into that kind of relationship. I said all this without anger and was shocked at myself even while the words were coming out of my mouth. Needless to say, he stopped calling me every day.

The next opportunity also came through dating and, once again, I confronted without anger. I began to notice that after confronting men I felt no remorse as I normally would have. There was no guilt about hurting their feelings.

The next evening Devin and Joysia informed me that my last implant was being removed that night, and, that I would be finished with recoding. My last implant was located in my neck. They also informed me the "Mouth of God" opening, located in this area, would be sealed and I would no longer have to worry about neck pain. This was August 21st. I was finally through recoding, and reconnection was next. The remainder of August and September was spent building a client base for my channeling business. I call it a business for that is what it is. Since I derive my income from channeling, I need to treat it with the same amount of attention and detail as a regular business, so Devin says.

During this time period, he was coaching me how to set up a metaphysical business in ways that are fun, profitable, and win/win for my clients and me. This we did. I went from one client a week at the end of recoding in August, to twelve and thirteen clients a week by the first of October. I accomplished this through networking.

Devin guided me to select and become involved in six local business networking groups. This was in addition to the CRN network Marla and I had developed, and that I still attended. It was fun to watch peoples' faces when I told them what I did for a living.

Everywhere I networked, I had small groups of people hovering around me asking me questions. They were genuinely interested, and I would have to say "no more," to be able to extract myself from the group. This required a leap of faith, financially. Devin explained to me in late July, that if I would let go of CRN and give it to Marla, I would be rewarded in many ways, including financial. I did, as of August 1st.

This would give me approximately $3,000. We had just mailed our quarterly billing in July and this represented the remaining balance of outstanding receivables. Everything else went to Marla. I had enough money to get me through August and September, but come October 1st, I was on my own.

In late September the money ran out, but, as I watched it run out, I saw my client base increase. October 1st, my rent was paid on time for the first time in months and October ended with my income being nearly double what I had been bringing home from CRN. Devin was right again. The leap of faith had really paid off.

Marla is taking CRN to new heights and I am so happy for her. Her financial worries are over and mine are too. The last eighteen months of financial hell were over. The company had been turned around by Devin, through his idea of changing the networking training from bi-monthly networking meetings to a six-week training class. Her classes are full and the memberships in the networks have nearly doubled in the two months since I left.

171

18

International Conference

The WE International Conference in Springfield, MO

Devin/Anu had channeled at the monthly channeling session in September that I would begin public speaking as the next piece of my contract very shortly. Once again, he was right. What I didn't know was that I had completed my major training and was ready for service. The WE Conference would provide the opportunity for my debut.

The WE (Walk-ins for Evolution) Conference in Springfield, on the weekend of October 27 and 28, was my first speaking engagement. WE is headed by a wonderful woman by the name of Liz Nelson. This was a regional conference, and I had gotten the information on this organization from Susie Konicov (Shalandrai), of *Connecting Link Magazine*.

I contacted Liz in September and shared my information. She was interested and asked if I would consider speaking at her regional conference in late October. I said I would think about it and get back to her. I wrote about it in my journal. One day I wrote that I had decided to go, and the next day I wrote that I had decided not to. Back and forth, back and forth I went.

On the 19th I had decided not to go. The next day Liz called and asked me again to consider speaking. All during this time Devin kept saying I needed to go, that there were people there I needed to meet. Finally I acquiesced. When Liz called again, I said yes.

While driving to Springfield for the conference, Devin asked me to pick up a piece of lapis lazuli. I didn't know how this would be accomplished. I wasn't aware at the time that someone would be selling stones at the conference, as this was my first metaphysical conference. I was only four years old, having walked in in June of 1992. I mentioned my concern to him about where to find it, and he said someone there would bring it to me. With that I forgot about it and continued my drive to Springfield.

I arrived just before the noon break. My talk was scheduled for 2:00 p.m. I was invited to lunch by a group of attendees. During lunch we took turns sharing our names and something about ourselves. I was the last one to share, and when I did, the people kept asking for more information. I said I would give it during my presentation. It was to be one of the three 20-minute break-out sessions that would be held before the main meeting reconvened at 2:30.

Nearly 20 people showed up for my presentation (there were only about 35 attendees at the conference). Devin was channeling the words as usual, but instead of discussing accelerated DNA recoding, I began with the history of the universe. When Liz came to tell me our time was up, the people were sitting on the edges of their chairs. Liz agreed to give me another time slot later in the evening at the close of the main meeting. About 20 people showed up for that session, too—this was very good attendance!

After the first presentation in the afternoon, one of the attendees, Clara Bo, came up to me and asked if I knew anything about the planet with the two suns. I said no, but commented that someone else had asked me the same question before. Later that day I remembered who it was. It was Rio, a client of mine from Charlotte, N.C. She had been referred by another client, Terry. That was the beginning of the Avyon information.

During one of the breaks, I met a wonderful lady, Kari Chapman, to whom I will always owe a debt of gratitude. Kari told me she had the sword Excalibur. I wanted to see it because I had always been a believer in the Camelot story. I had just read the book *The Mists of Avalon* weeks earlier.

Later that evening, before my presentation, I went up to her room, and she showed me the sword. I held it in my hand and felt waves of energy surging through me. At that moment Devin said to me,

"Hold the wand against the crystal cluster in your shoulder." I asked Kari to do this for me.

I felt the cluster begin to buzz when the wand touched it. There was a warm tingling sensation as well. Kari moved the wand in a circle, then changed directions and moved around the cluster again—a very simple, unceremonious ceremony. Just the way Devin prefers them. That night the information of Avyon began being downloaded to me. I remember having a fever and pressure around the top of my head, but I wasn't sick. I was simply vibrating at a higher frequency. This continued for the remainder of the conference.

On Sunday morning I went back to Kari's room to tell her about the experience I had with two big etheric cats. They were the size of leopards and had been playing in my room during the night. I was sleeping on the sofa and the cats had come in through the open window nearby. Two of the other women with whom I was sharing the suite had also seen them. I quickly realized that it was due to the picture I had with me of Joysia and Anu. Devin said to cover the picture and the cats would leave. It seems they had come in because they were drawn by Joysia's Feline energy.

As I entered Kari's room, I noticed all the rocks and stones laid out for sale on her bed. Suddenly, I remembered Devin talking to me about picking up a piece of Lapis, so I asked her if she had any. She said she did bring three pieces and proceeded to show them to me.

Two of them were regular unpolished pieces and one was in the shape of a ball. I held each of them, and when I picked up the ball I knew I had the right one. When I showed her the one I had chosen, she looked at me with surprise and told me that she had decided to leave that one behind, but at the last minute, changed her mind. She said Spirit had told her that she needed to take it, because it belonged to someone who would attend the conference. The Lapis ball was very old and had come from Afghanistan. It even had papers! I had no idea that rocks had papers! So I purchased the Lapis ball for $8.

In time Devin would share more about this little piece of Lapis. It seems it will be instrumental in accessing the records stored in the paw of the Sphinx.

The next day, Sunday, I was scheduled to leave no later than 3:00 p.m. I had to get back to Kansas City to do the monthly channeling session at Ann's house. I had told Liz that I did not want to do any readings, yet two people had booked appointments, and Devin said it would be wise to honor the bookings.

The first reading was for Clara Bo. Her initial question was, "Tell me the name of the planet with the two suns. " The name that came through was Avyon. Before I left the conference, I received more information on Avyon. And I received still more during the following three weeks with the majority coming during the next three days.

Sunday evening I returned to Kansas City, one hour before I was to channel for the monthly session. I drove straight to Ann's house. Another car arrived just as I was walking up the walkway to the front door. I briefly noticed it but promptly forgot about it as soon as I walked in the front door.

Minutes later the doorbell rang and there were Susie and Barrie Konicov of *Connecting Link* magazine! During my drive home from Springfield, I had forgotten that they might be there. They had called earlier in the week and said they were heading my way and might arrive in time for the session.

That night Devin/Anu talked a lot about Avyon. It was a wonderful session. He reminded everyone of the predictions he had made back in June; all of them had come to pass. People left the session that night with a new sense of conviction and belief that all this was real.

After the session, Susie, Barrie, and I decided we wanted to talk for a while, so they followed me home in their car. Later, when they were ready to find a hotel for the night, I invited them to stay at my place. They lived with me for several months and we were like a small family.

It seems Devin and the Council used their motor coach, Freddie, as a tool to get them to Kansas City. Once here, Freddie broke down and it took nearly three weeks to fix her. During that time we became acquainted and found out we were 9th-dimensional brothers and sisters.

Devin and Joysia helped Susie and Barrie begin their recoding the day after they arrived. I channeled Devin and Joysia for them and a plan was laid out for them to follow. I remember wondering how they were going to get through recoding in a matter of a few weeks when it had taken me eight months. I knew that the Council was planning on accelerating the process, but how were they going to speed it up that much?

Once again, Devin knew what he was doing. He sent them to three people for different cleansing and preparation techniques, and each also completed a liver cleanse. Barrie downloaded his 9th-dimensional oversoul, Kavantai, on the first night of his recoding process. He was also given the correct spelling of his name.

He and Susie had known their 9th-dimensional names before this (Kavantai, Shalanah), having received them through a reading years before, though they were not given the correct spellings at the time. We were also being given information of our 9th-dimensional family ties.

Over the next three weeks Kavantai, Shalandrai ("Shalanah" evolving into "Shalandrai") and I continued receiving information and training to prepare us for the next big step, The Star Vision conference in Estes Park, Colorado, on November 9th.

We found out about the 90-member oversoul cluster, made up of Felines and Carians, that had come to this universe, by invitation of the Founders. We learned they had birthed into the Amelius Line, The Royal House of Avyon, on the Lyran planet Avyon, at the 9th-dimensional level. What a shock to find out that Devin was the Patriarch of the Amelius line at the time they left the Lyran Avyon, and that we were his brothers and sisters! We were all members of that 90-member oversoul cluster. Finally, all the pieces of the puzzle were coming together.

We learned that this group, the 90-member oversoul cluster, also oversaw the Founders' Polarity Integration Game for this galaxy. And that the story of Abraham was an Earthly rendition of Devin's story. We also found out that Grandfather Two Moons was Zephrin, and Zephrin was the 11th-dimensional spokesperson for the

Founders. He sat on the Council of Twelve. What an incredible three weeks!

Next, we were given the 13th-dimensional Formula of Compassion for permanently releasing negativity from the body. Then we learned about Marduk and the Dark Forces and the need to integrate him so he could begin working with us to end this Polarity Integration Game.

After the three of us had integrated and released Marduk, Devin sent us to the woods to perform a simple ceremony. We were asked to take a large generator crystal, my dog, and my child, so they could act as proxies for the animals and the children.

We drove to the park and walked up a trail into the woods. Devin guided us to a site by an old oak tree. We sat in a circle and placed the crystal in the middle. Devin channeled the words and we repeated them. Next, we turned the crystal upside down and repeated the same words. The purpose for the ceremony was to proxy for the galaxy, to activate a dormant code within the bodies of all beings in this galaxy, both Human and Reptile. Devin had me call it the Christ Consciousness Code. The proxy was performed by Kavantai, the Galactic Proxy and Devin's brother.

I remember Kavantai saying as we were leaving the woods, "Well, that sure was an unceremonious ceremony, no pomp, no anything!" Devin had me give him a message, "That simple ceremony will have far-reaching positive repercussions," and it did. A week later we left for Estes Park and the repercussions became glaringly apparent. Devin believes in simple ceremonies with *big* results!

Two other Avyonian family members had begun the recoding process before we left. They were Terry and Maggie. Both had been coming to me as clients, and I was using the vow-rescinding regression process on both, to clear major blocks.

The difference was dramatic. I did not know at the time that the process included the 13th-dimensional Unconditional Love Formula. Zephrin and Devin had not given it to me as a formula at that time, but I had been working with it and learning about its power since early September.

The change in Maggie was the most amazing. Maggie had been coming to me to work on a powerlessness block. We had released it and I had not seen her for a month. When I saw her at the monthly channeling session on October 27th, I was shocked. Instead of a woman who frowned and was depressed, she was a happy, vibrant, laughing individual. It was my first experience of witnessing the power of the process Zephrin was teaching me. From that point on, I was convinced of the power of the vow-rescinding process.

On Wednesday, before we left for Estes Park, another member of the 9th-dimensional Avyon family arrived, Lois Hollis. We had met over the phone. She was a *Connecting Link* subscriber. Lois had ordered a set of the DNA Recoding Tapes we had begun distributing through the *Connecting Link*.

After listening to a few sentences, she picked up the phone and called Susie to get my number. A voice within told her this was her family and she didn't waste any time in reconnecting. Lois too, began recoding, and, like Susie, Barrie, Terry and Maggie, she would complete the process in Estes Park.

19

Star Vision Conference

Star Vision Conference - Estes Park, Colorado
November 7-11, 1996

On November 6th, the night before we left, Devin and Zephrin had us convene a meeting for a small group of people. These were individuals who Devin said were members of the Avyonian family and the 90-member oversoul cluster.

Zephrin channeled that we would go to Estes Park and speak. At the time we were not on the speaker list. He said, "Many will come, seeking the knowledge you bring." He said that we would hold a workshop as well. You can imagine how I felt. Here I was telling the people all this would happen. What if it didn't? Well, it did, every bit of it, and even better than I could have imagined.

We left early in the morning, Versarai (Terry), Kavantai (Barrie), Shalandrai (Susie), Shasarai (Lois), and I. Maggie, who would receive her name when we returned, rode with others and joined us later. This trip would change all our lives. The things that were important before would not be after, and those that weren't important before, would be, now.

We were given a one-hour slot for a presentation in the main auditorium. This occurred Friday morning. Devin, once again, guided me, and 8:30 a.m. on November 9th, was Zero Point, as Devin called it. I had not heard that term before and found it strange. During the four days of the conference I heard it over and over, always connected to a man named Gregg Braden. Gregg wrote the book, *Awakening to Zero Point*.

After the presentation, we were told that we could hold a free workshop. We scheduled it for Sunday morning at 10:00 a.m. Close to a hundred people signed up for it. We tried to tape it, but the recorder wouldn't tape. Later on, the person in charge of taping came up to me to tell me that the recorder began working again soon after the workshop was over.

Devin said he didn't want it taped, but once again, I blew him off. Devin always has his way in the end. After all, he's in charge of this program and knows better than I what is appropriate in the moment. Once again, Devin was right.

The Avyonians did come forward. People came to me, saying the information I had shared was what they were looking for and the reason they had come to the conference. I had readings booked, and with the exception of one, they all were for people who were Avyonians. At each reading Devin assigned a role to the person in the forthcoming work. Everyone loved the role Devin gave them, and I found that each one had spent many years in training for that role. As with me, the Council had been working with them.

During the entire conference the undercurrents of emotions were very strange. Devin said the portal in Estes Park had been opened and the Forces of Light and Dark had entered en masse. He said the final act of the Galactic Polarity Integration Game had commenced.

The starting point was my presentation, when I gave the audience the 13th-dimensional Formula of Compassion to integrate the Reptiles and their representatives who had come for the conference and for the Game. From that point on, strange things happened. People were talking about the undercurrents and what was really going on.

On Monday morning, I awoke with Devin telling me that he wanted me to gather the group and go to the creek in the woods. We were in the YMCA compound, and there were woods all around. So I did.

Devin guided me to the spot. It was a beautiful setting, with ice covering the sides of the creek. Then Devin guided me to have each person make a vow to serve the Light. He said to tell them not to make the vow if they did not feel they could honor it, and if they didn't make the vow there would be no judgment. The vow was merely a recommitment to the contract they had made before entering this lifetime.

Everyone went off and made their vow in their own way. Then we gathered together again. Devin channeled and I spoke. I told everyone that it was time we integrated the Dark Forces, that they were souls playing a role just like us. I also reminded them that the Dark roles are harder to play than the Light roles.

I asked them to pray for Marduk—his has been the hardest role, and he wants to come home. Then Devin had me tell them to use the 13th-dimensional Formula of Compassion to resolve any conflicts that come up. He said we are Avyonians and, therefore, representatives of the Light, and here to share the 13th-dimensional Formula of Compassion. After that, we left the creek and walked back up the hill to the road.

As we walked back to the camp, one person in the group happened to look up, and over the top of the mountain was a huge cloud in the form of a ship. Devin said it was, indeed, a ship and our confirmation that all we were experiencing was real. As I said earlier, Devin is notorious for small, simple ceremonies with *big* confirmations.

Joysia completed the recoding and reconnection process on Kavantai, Shalandrai, Sharsarai, Versarai, and Maggie. All except Maggie had their Oversouls downloaded. Maggie would gain permanent access to her Oversoul since her contract was a little different than the others. Devin explained that the majority of individuals recoding would have the same Oversoul arrangement as Maggie.

Devin said their recoding data was needed, as it would provide critical information for Joysia and the other genetic engineers. It seems the frequency of the Estes Park vortex is higher than the vortex in Kansas City, the high heart of the Dove. This data, he said, would be used by the Council and the Felines to shorten the recoding and reconnection process. Once it was analyzed, Joysia would give me an accelerated version of the process.

On Tuesday morning we left the conference and drove home. It had been an eventful four days. It felt as if we had been in a time warp and now were returning to real time and the real world. During the drive home we received even more information. It took me over a week to process it all.

Two weeks after Estes Park, we received information that scientists had spotted Nibiru the night of November 14/15. Now, the scientists do not know it was Nibiru, but Devin says it was. It was spotted next to the comet, Hale-Bopp. Devin says the comet is Northwind, the satellite that always precedes Nibiru on its 3600-year orbit through our galaxy. So much happening in so little time! It seems things are occurring every day. Devin said it would be like this, and once again, he was right.

Today is December 1, 1996. Devin says I am to end the book here. It is the anniversary of my awakening on December 1, 1995. This evening my ex-husband, Rick, called and wanted to talk to me. I went to his house, and he began to tell me information he had received during a meditation. I never knew he meditated!

The information was about Avyon. Rick, my former husband and father of our daughter, Danielle, is one of us! I would never have guessed it. What a shock. Now my daughter will not be torn between her father's teachings and mine. They will both be the same. Rick will be working with me and traveling. Our daughter will go with us. What a gift!

Devin told me that when you release someone through the 13th-dimensional Formula of Compassion, you also release them from playing the dark role in your life. This allows them to continue their path and come back around, but this time in a different way. This happened with Rick and me. Devin said I would be rewarded, both here on Earth and when I return home. The return of loved ones, for me, is the greatest gift of all.

20

Life After DNA Recoding, Reconnection, and Activation

Life after recoding is different. I find I do hold and use my power in ways I only used to dream of. My ability to remove self-created blocks is quite powerful, and my ability to manifest is greatly improved. My self-esteem is much higher, and my love for Earth and mankind is stronger. I like being here for the first time since I walked in four years ago. I love my contract, and I am experiencing a joy in fulfilling it that words cannot adequately describe. I am truly happy now.

The most remarkable change is the way that I interact with others. I can so clearly see where they are coming from. When my feelings get hurt, and they still do sometimes, I use the Formula, and then I am able to have compassion towards the person instead of anger. I realize they are merely playing a role for me so I can see an aspect of myself that needs adjusting.

There is still a place for anger, and I experience my share of it, but instead of allowing myself to internalize it, I release it. I don't harbor anger anymore. I feel much lighter because of it. I see how dense anger is and how it literally weighs us down, when held in our bodies.

Depression, which used to haunt me regularly, has not come to visit in months. I am also more social than I used to be. I find I desire being with people and find I yearn for a relationship with a man. I used to find men a waste of my time and enjoyed being single. Now

I long to be married and to settle down. I would say this is one of the biggest changes of all!

Speaking of being married, I have not met my lifemate yet. I know he is the oversoul of Enki and, therefore, works with the ancient sciences. As Enki's oversoul, he is intimately aware of the knowledge and purposes of the Egyptian Temples of Initiation and the Brotherhood of the Snake that Enki started.

Devin said I would meet him after the book was published. I guess that means any day now. Devin says that when we meet, we will know each other instantly. We will combine our work and when that happens we will be able to give mankind the complete picture.

I will give the spiritual and historic half, and he will give the scientific, mathematical half. Events will continue to move rapidly from that point forward to the shift of Earth to her 5th-dimensional state. I feel life is truly wonderful. I wish the same for you.

Postscript

It has been two years since I sat down and wrote *We are the Nibiruans*. *So* much has happened that I felt a postscript was needed to bring you up to date.

The year 1997 was a year filled with change. The Nibiruan Council organization was established along with a physical office and training center for workshops. My good friends, Malarai (Dermot Kerin) and Versarai (Terry Spears) joined me as partners to bring the 3D portion of the Nibiruan Council into being.

In March we published our first newsletter, the *Nibiruan Council News* with Susie Konicov of the Council as editor. The response was tremendous. Then Susie and her husband, Barry, moved on to their next step.

We began holding workshops in March and many came knowing that the multi-dimensional information was part of the knowledge they needed to complete their earthly assignments. Many of the people were walk-ins and starseeds and some became coaches, teaching and sharing the Formula of Compassion with those who came in search of assistance.

I began my first-ever speaking tour in September, 1997, spreading the word that the multi-dimensional tools of integration were now on Earth as well as an accelerated process for recoding our DNA. Many heard the call and came forward to receive what they needed.

In late 1997, I met Snake Dancer (he later gave me his 9D name, which is Jehowah). His Earth name is John Starr. Snake Dancer arrived on Earth as a walk-in shortly after we got together in early December. He and John exchanged places on December 10, 1997. We married on January 13, 1998. He told me that when he arrived we would move quickly because we had much to accomplish in a short time.

In 1998, John (now Jonathan; he changed his name) and I moved to his hometown of Los Angeles. This was a difficult decision for me, as my daughter would not be coming with me. Danielle would stay in Kansas City with her father and come to L.A. for extended visits.

We closed down the Council headquarters and moved everything to Los Angeles. Versarai and Malarai, two beloved friends that had been with me almost from the beginning, moved to Cedar Rapids, Iowa.

In August 1998, I completed my second book for the Nibiruan Council. It is called *Bridge of Reunion, A Lightworker's Guide to Mending Broken Family Ties.* It was hard coming up with a subtitle because the book was about so many things. It is the story of my and Joscelyn's spiritual path going all the way back to her early teens. It is also about the lessons that we must learn as part of the journey of the spiritual path.

In September 1998, the dimensions split as the sun's corona entered the photon band. We were now living in two dimensions simultaneously. People everywhere were feeling the physical and emotional effects as relationships broke down, computers and the internet went off-line, and communication between individuals became fractured. Based on people's consciousness, they, in some cases, had to communicate across a dimensional bridge. This further stressed already strained relationships on both the personal and family fronts.

By October, I realized that many were needing help with using the Formula of Compassion to release certain types of emotional blocks. The workshops helped a lot, but the problem was that people were getting stuck once they returned home to carry on with their day-to-day lives. The Council, as always, was aware that this would happen and, therefore, they asked me to write the tools down as an aid to clearing emotional blocks and attaining real compassion. As a result, I began writing *The Multi-Dimensional Keys of Compassion.*

The Keys (my nickname for them) developed as a set of seven booklets that gave detailed information on four additional tools to use in clearing stubborn emotional blocks. The four additional tools, The Compassion Key, The Soul/Ego/Self Partnership Key, The Open Door Key, and The Hold Onto Nothing Key, are tools that helped in completing the nine steps of the Formula of Compassion. They are based on painful real-life struggles that I had endured in achieving multi-dimensionality and compassion myself. They unfolded as warm, endearing and painfully honest stories of the joys and sorrows that one experiences on the journey to understanding, self-love and acceptance.

Although two of the Keys, The Formula of Compassion and The Compassion Key, were written before I left Kansas City in July, 1998, they had never been written down in such complete detail until I wrote all seven of them in October. Now mankind had all seven of the higher dimensional tools for achieving integration of the Light and the Dark in one easy-to-read set of booklets. And all but one had a companion audiotape. I felt that we were really making progress now.

As January 1998 rolled around, I found myself doing a major overhaul on our website which had not been updated for a year. At the time, I couldn't understand why I was taking on such a monumental job, but now I know. It was in preparation for the next phase of the work we were here to do. So much is happening so quickly now.

As of today, March 1, I see that I will be doing more work from home yet, at the same time, reaching millions of people. It is time to prepare mankind for the events that are happening right now and that will affect everyone's future. They need to be prepared for the inevitable uncloaking of Nibiru.

Through our massive website and the radio interviews that I continue to do around the world, I am reaching the masses and beginning to accomplish the job of preparing them. What happens this year will set the stage for the next 10 years

1999 is to be a year of plans finally coming together, of projects coming to fruition, of people uniting and really getting on with their assignments and purpose for being here. I am so grateful to be a part of such an unprecedented time in our galactic and universal history.

Jelaila Starr
Los Angeles, CA
March 1, 1999

Jelaila Starr

Photograph by Steve Dorey

Mr. Dorey is a specialist in intuitive and environmental portraiture.
He is located in Redondo Beach, California at 310.375.2375.

Biography

Jelaila is a Nibiruan Council's messenger, but more importantly she is a mother, wife, teacher, entrepreneur and visionary. After her spiritual awakening experience in 1992, Jelaila worked to heal the emotional wounds of her past and thus began the Spiritual Path.

In December, 1995, Jelaila was awakened to the knowledge of her walk-in. The following January Jelaila began the first phase of her mission, DNA Recoding.

In July 1996, after six successful years with CRN (Creative Referral Networks, Inc., a company she started in 1992, Jelaila felt a calling to change her career focus to a new spiritual and humanitarian field which brought her into alignment with her assignment as a galactic messenger.

In January 1997, Jelaila completed her initial training as a galactic messenger and founded The Nibiruan Council, named after the 12th planet of our solar system, Nibiru. The Nibiruan Council was created to be a vehicle through which Jelaila could publish her writings as a spiritual visionary and messenger of the higher knowledge of our origins as a race, as well as the bigger picture of our galactic and universal history.

This knowledge coupled with Jelaila's simple, uncomplicated and down-to-earth approach, enables her to reach across the boundaries of belief systems that separate our world, to provide insight, understanding and hope and, more importantly, a concrete formula for self empowerment.

Jelaila's direct and open manner is refreshing and welcome in this time of complex and often complicated approaches to understanding our world. At this time when many speak of doom and catastrophic change, Jelaila's is a voice of calm, reason and hope.

Jelaila is living proof that you can be spiritual, happy, healthy and financially successful at the same time. Through her own example and experiences, she provides an alternate way out of the earth systems, while still taking part in the good they have to offer.

Jelaila has authored two books for the Nibiruan Council in addition to other Council information available through booklets, audio tapes, videos and workshops.

Jelaila lives with her husband Jonathan, and 4 cats, Biijai, Rai, Lily and Ginko. She can be contacted through the Nibiruan Council's website at www.NibiruanCouncil.com.

Glossary of Terms

Aln - The planetary birthplace of the Reptiles. Located in the constellation of Orion.

Amelius - The first aspect or soul fragment Sananda created. Amelius was the first soul to incarnate into a human body on the Lyran Avyon.

Amelius Line - The soul and bloodline of Amelius. Later became known as the Royal House of Avyon. The Amelius line is the pure line of the Humans and used for seeding purposes by the Felines.

Aquatic Sirians - The Etheric Sirians who chose to mutate back into aquatic mammals. They are the whales, dolphins, and merpeople.

Avyon - The planetary birthplace of the Humans. Located in the constellation of Lyra.

Battlestar - A star that has been equipped for military and peacekeeping duties. A battlestar is a galactic police force.

Carians - The bird people. The Carians came here from another universe to construct the star gates, dimensions, planetary grids, portals, and vortices. They are the magnetic engineers for Divine Creator.

Carian Magnetics Engineers - Carians who perform the work of creating the dimensions, stargates, portals, and grids for universes, constellations, solar systems, and planets.

Christos Office - An office of Etheric Sirians, headed by Sananda, attached to Earth's planetary Spiritual Hierarchy. They coordinate the fulfillment of the Etheric Sirians' Divine Plan on Earth. The Christos Office also works in conjunction with the Nibiruan Council to fulfill the Divine Plans of those races and civilizations involved with the Etheric Sirians' Divine Plan.

Christos Sirians - Those Etheric Sirians who form the Christos Office.

Council of Nine - The group of nine First-Source Souls who manage our universe.

Council of Twelve - A group of beings who are the first soul fragments of the Council of Nine. They carry out the directives and act as spokespeople for the Council of Nine.

Council of Twenty Four - Game Overseers. Also responsible for creating all the souls in this universe, with the exception of the 90 Game Engineers and the fragments. Also responsible for coordinating communications between the Angelic Realms and the Incarnate Dimensions.

Earth Sirians - Etheric Sirians who became stuck in the animal incarnation cycle.

Etheric Sirians - Humans from the Lyran planets of Avyon and Avalon who chose an etheric state of existence and are feminine-polarized. Heirs of Earth. They divided into three groups, Christos Sirians, Aquatic Sirians and Earth Sirians, to fulfill their Divine Plan and to play their parts in the Galactic Polarity Integration Game.

Felines - First-Source Souls from a neighboring universe who have completed their Polarity Integration Game. They are responsible for the creation of all life forms in this universe. The Master Geneticists for Divine Creator.

Feline Construction Engineers - Felines that create planets, constellations and star systems.

Feline Genetics Engineers - Felines that create lifeforms for planets and stars, including physical vehicles for souls.

Firmament - The three-mile thick band of moisture in the atmosphere surrounding Earth.

Founders - The Council of Nine

Game Engineers - The 90 Felines and Carians who came to our universe to set up and manage the chosen Universal Game of Polarity Integration.

Galactic Federation - The multi-dimensional organization of Councils that represent the many races and civilizations in our galaxy. Responsible for maintaining positive communications and peaceful relations. Also responsible for coordinating efforts between the Angelic Realms and Incarnate Dimensions to fulfill the Divine Plans of all races and civilizations involved in the Galactic Polarity Integration Game.

House of Aln - Royal line of the Reptiles. Descended from Lucifer, one of the Council of Nine.

House of Avyon - Royal line of the Humans. Descended from Sananda, one of the Council of Nine.

Hybornea - First Human colony on Earth. Destroyed by the Reptiles during the Galactic War.

Land Guardians - Guardian race of a planet. Must be in physical form to care for and protect the planet.

Lapis Lazuli - The stone that holds the frequency of Avyon. Worn by all Avyonians.

Lemuria - The second colony established on Earth. Consisted of Humans from other galactic colonies who chose the Matriarchal form of society. Later became the Mother Empire to the new colonies of Yu and Atlantis.

MEs - Encoded crystals shaped like a wand or ball. Contains information and programming to operate things such as a library, a star ship, a planet, or a star constellation or cluster.

Nibiruan Council - The Council that represents the entire Royal Houses of Avyon and Aln, incarnate in all dimensions. Called the Nibiruan Council due to the transfer of both Royal Houses to Nibiru after the destruction of the planets, Avyon and Aln.

Picts - A group of Lemurians that left Lemuria and settled in Northwestern Europe.

Rama Empire - A group from the Yu Empire that moved underground and later surfaced as the Rama Empire.

Seeding - The distribution and placement of biological and genetic materials on a given planetary surface.

Sirius A - Located in the Sirius star system. The home of the Felines.

Sirius B - Located in the Sirius star system. The second home of the Lyran Humans.

Sirius C - Located in the Sirius star system. Used for warehousing of materials and laboratory facilities by the Felines.

The Things - Part animal and part Etheric Sirian.

Twin Flames - Two halves of the same soul.

Yu Empire - Located in the area of China. Home of the Yellow race. Colonized by Ashan, a cousin of Anu.

Summary

All of the help and information to recode and regain your multi-dimensionality is available now!

This is a self-help program. You can do this yourself. Help is available for those who want and need help along the way.

Book One has the road map of the changes that all of humanity is going through to ascend. This is also available on Tape 4, with examples. The Formula of Compassion, how you can get through the changes, is also in Book One, and on Tape 5. Most people find this process easier when they have the Book and these two tapes.

The Liver Cleanse suggested by Joysia helps to jump start your process. We all store anger in our liver. When we can express emotions, we can begin to release them. You can peel four to six layers off your emotional blocks with this cleanse, to get to your stuffed anger, begin to express it, and release it.

Two supplemental tools to help you through your process:

- **The Compassion Key** to help you unlock your heart, for those of you who shut it down, when it just got too painful to feel. On Tape 16 and in the Compassion Key Booklet.

- **The Soul/Ego Partnership** to help the ego partner with the soul, who can see your life blueprint and create multi-dimensionally. It's the answer when your ego blocks you in fear, or just doesn't know where to go and what to do anymore. On Tape 19.

These are the basic tools and the road map to get started. They are available through the website at www.nibiruancouncil.com.

Workshops are offered that will help you learn these tools more quickly and fully, and peel off many layers of your emotional blocks, with the help of trained coaches assisting. The website can provide you with a catalog and a list of coaches. The coaches will be happy to hear from you and help with your questions.

The Nibiruan Council

For more information on this book, or to inquire about workshops, seminars, and sessions with Jelaila, or the Galactic Counselors, please contact:

Jelaila Starr
The Nibiruan Council
Website: www.nibiruancouncil.com
Email: jelaila@nibiruancouncil.com

To obtain additional copies of this book, and additional books, tapes, and videos of higher dimensional information related to:

- The DNA Recoding Process
- Universal/Earth History
- Walk-in/Starseed Information
- Higher Dimensional Children
- Earth's Progress towards Ascension
- Coming Earth Changes
- Current Events
- Multi-dimensional knowledge
- The Galactic Federation and Nibiru
- Much, much more

Please contact us at the address currently shown on our website for a complete catalog, or you can order from the Nibiruan Council's Online Store.

The Nibiruan Council Mailing List

Now you can get the latest information from the 9D Nibiruan Council regarding earth changes and events, new products and information, along with up to date workshop schedules and more by joining the Nibiruan Council Mailing List.

Website: www.nibiruancouncil.com
Email: info@nibiruancouncil.com

The Multi-Dimensional Keys of Compassion

The Sequel to Part 3 of

WE ARE THE NIBIRUANS

You've been asking for pocket sized versions of the multi-dimensional Keys, and now they're available in booklet form. Each of the booklets give a strong narrative on how the key came to be, as well as how to use them.

The Multi-dimensional Keys of Compassion are the sequel to Part 3 of We are the Nibiruans, Return of the 12th Planet.

- **The Formula of Compassion** explains the nine step process of healing and releasing through the high heart.

- **The Compassion Key** provides the steps to release from blame those with whom we have the most painful emotional blocks, especially parents and other authority figures.

- **The Soul/Ego/Self Partnership** explains the roles of each of your selves (Soul & Ego) and how they work with you (Self) on your life blueprint.

- **The Open Door Key** describes how you can create a loving partnership with Soul and Ego so you can work together towards a common goal and happiness.

- **The Hold on to Nothing Key** shares the higher dimensional tool and steps to use to integrate our deepest fears. Releasing the pain and associated problems created by our deepest fears is the pathway to a happier, satisfying multi-dimensional life. Funny and full of wisdom.

- **Dancing with the Dark** provides the steps and awareness needed to integrate the dark, the other polarity and our greatest teacher.

- **The Agreements Key** provides the tool needed to create safe, enjoyable and satisfying relationships as well as heal the wounds that unbalance existing ones.

The Nibiruan Council's Multi-dimensional Workshops

Join *Jelaila Starr* and the Galactic Counselors for a fun and enlightening event. Change and enhance your life by living in the Consciousness of Multi-dimensionality.

The following workshops are part of the Galactic Training Series, a series of four workshops designed to return you to your natural multi-dimensional state, while at the same time, preparing you for galactic citizenship.

Galactic Training Level One - Universal History/ Galactic Politics

Open your mind and expand your consciousness as we take you on a journey through our past. To know where we are going we must know from whence we came. This workshop includes a study of galactic politics, the ancient history of the reptilian/human conflict and the state of our current situation as we move towards galactic citizenship. A powerful workshop designed for those ready to embrace and live life multi-dimensionally.

2-day workshop

Galactic Training Level Two - Ascension Tools I

Following on the heals of Level One, we explain the Accelerated DNA Recoding Process and the multi-dimensional tools given to complete it. This is an information-packed workshop that covers the details of DNA Recoding and the basics of the *Multi-dimensional Keys of Compassion*, the unique higher dimensional ascension tools.

2-day workshop

Galactic Training Level Three - Ascension Tools II

There is so much to learn about the multi-dimensional tools used for ascension. This workshop picks up where Level Two ends and focuses on emotional clearing techniques, Inner Child work and self-empowerment through agreements. These skills are a must for those choosing to enter the galactic neighborhood.

2-day workshop

Galactic Training Level Four - Galactic Mastery

This workshop completes the work of the first three levels and provides the final teachings that complete the training of a future galactic citizen. Included are the multi-dimensional knowledge and techniques used to reconnect you with the higher realms, both incarnate and angelic. This knowledge, previously known to only a few on Earth, is now being returned to all Mankind.

2-day workshop

For a complete listing of all of our workshops, visit the Workshop section of our website at www.nibiruancouncil.com.

NC Forums and Chat rooms—Support for DNA Recoding and Emotional Clearing is available now!

DNA Recoding, though it provides such wonderful benefits can be very challenging while going through it. If you would like assistance through the ups and downs of DNA Recoding you will find it through the *Nibiruan Council Forums*. The Forums are moderated by Galactic Counselors, individuals who are trained to provide counseling to those in DNA Recoding as well as assist in guide communication, relationship counseling and a host of other unique skills.

In the NC Forums you will not only find great support but you'll meet people like you from around the world who are going through similar experiences. The NC Forums are our online support group for DNA Recoders.

The *NC Chat Rooms* are home to the Monday Night Workshop, a free mini workshop by chat facilitated by the Galactic Counselors. Discussed are a variety of topics ranging from ascension to universal history. An exciting and educational experience awaits you! Check the *Event Calendar* on our website for the current schedule.

The Nibiruan Council
Email: info@NibiruanCouncil.com
Website: www.NibiruanCouncil.com

(Our current address and phone numbers are listed on the Nibiruan Council website)

p.64   Nibu — Siki in Pln